T0257795

Biomedical Tissue Culture Handbook

Biomedical Tissue Culture Handbook

Edited by **Hugh Waddell**

LANRYE
INTERNATIONAL

New Jersey

Published by Clanrye International,
55 Van Reypen Street,
Jersey City, NJ 07306, USA
www.clanryeinternational.com

Biomedical Tissue Culture Handbook
Edited by Hugh Waddell

International Standard Book Number: 978-1-63240-087-1 (Hardback)

Printed in the United States of America.

Contents

Permissions

List of Contributors

Preface

This book focuses on providing information regarding biomedical tissue culture. It deals with the science of tissue culture highlighting nearly every major aspect of it. Among the many topics discussed in this book are; development of cultural techniques to produce neuron syncytial connections, multinucleated cells, method of regenerating cartilage tissue, improvised utilization of cell culture for virus isolation, cultivation of placenta derived cells, cells with properties of stem cells and cultures to study the pathogenicity of enteropathogenic bacteria.

This book is a result of research of several months to collate the most relevant data in the field.

When I was approached with the idea of this book and the proposal to edit it, I was overwhelmed. It gave me an opportunity to reach out to all those who share a common interest with me in this field. I had 3 main parameters for editing this text:

1. Accuracy – The data and information provided in this book should be up-to-date and valuable to the readers.
2. Structure – The data must be presented in a structured format for easy understanding and better grasping of the readers.
3. Universal Approach – This book not only targets students but also experts and innovators in the field, thus my aim was to present topics which are of use to all.

Thus, it took me a couple of months to finish the editing of this book.

I would like to make a special mention of my publisher who considered me worthy of this opportunity and also supported me throughout the editing process. I would also like to thank the editing team at the back-end who extended their help whenever required.

Editor

General Characteristics and Culture Conditions

Culture Conditions and Types of Growth Media for Mammalian Cells

Zhanqiu Yang and Hai-Rong Xiong

Additional information is available at the end of the chapter

1. Introduction

1.1. Basic requirement for culture medium

1.1.1. Nutritional components

Cells need the basic nutritional conditions to grow *in vitro*, including:

1. Amino acid

Amino acid is the raw material for the cell to synthesize protein. All the cells need twelve essential amino-acids: arginine, cystine, isoleucine, leucine, lysine, methionine, phenylalanine, threonine, tryptophan, histidine, tyrosine and valine, which are L-amino acids. Furthermore, glutamine is another component playing important role in the cell metabolism process. The nitrogen contained in glutamine is not only the source of purine and pyrimidine of nucleic acid, but also the essential material for the synthesis of the Tri-, bi-, mono-phosphate acid glycosides.

2. Monosaccharide

Cultured cells use aerobic glycolysis and anaerobic glycolysis of hexose as main energy source. In addition, hexose is used for the synthesis of some amino acid, fat and nucleic acid. Cell absorptive capacity varies among different monosaccharides, with the highest for glucose and the lowest for galactose.

3. Vitamin

Vitamins mainly act as coenzymes or prothetic groups in cell metabolism processes. Biotin, folate, nicotinamide, pantothenic acid, pyridoxine, riboflavin, thiamine and vitamin B12 are common component in culture medium.

4.   Inorganic ion and trace element

Besides some basic elements (including sodium, potassium, calcium, magnesium, nitrogen and phosphorus), cell growth needs some trace elements, such as molybdenum, vanadium, iron, zinc and selenium, copper, manganese.

1.2. Somatomedin and hormones

Cells grown *in vivo* are always regulated by somatomedin and hormones. Many researches demonstrate that various somatomedin and hormones are very important to maintain cell function and status (differentiated or undifferentiated). Some hormones have promoting growth effects on different cell typs. For instance, insulin can promote the use of glucose and amino acids in the cell. Some hormones are cell-type specific, as hydrocortisone that can promote the growth of epidermal cells and prolaction that induces the proliferation of mammary epithelial cell.

1.3. Osmotic pressure

Cells need an isotonic environment and human plasma osmotic pressure is about 290 mOsm/kg, which is thought to be ideal osmotic pressure to culture human cells. Mouse plasma osmotic pressure is about 320 mOsm/kg. Osmotic pressure of 260-320 mOsm/kg fits for most mammalian cells.

1.4. pH

The suitable pH for most cells is 7.2-7.4; otherwise it will produce harmful effects. The culture medium should have some buffer capacity. The main substance causing pH changes is CO_2 produced in cell metabolism process. In an airtight environment, CO_2 can combine H_2O_2 to produce carbonic acid and thus reduce the pH value of the medium. Synthesized medium employs $NaHCO_3$-CO_2 buffer system to solve this problem. In the buffer system, the boost in [H+] increases the reaction rate H+ + salt => weak acid and takes some H+ out of circulation. It is based on the constant equilibrium.

$$NaHCO3 + H_2O \longleftrightarrow Na^+ + HO^- + H_2O + CO_2 \uparrow$$

2. Natural medium

Natural medium is described as animal body fluids or medium of tissue extraction, including plasma, serum, lymph, chicken embryos leaching solution. Natural medium contains rich nutrients, various somatomedin and hormones, similar osmotic pressure and pH to body environment. As this medium has a very complicated production process and big batch-to-batch variation, the medium is gradually replaced by the synthetic medium. Today, serum is still the widely used natural medium.

2.1. Serum

2.1.1. Types of serum

Serum can derive from different animals. Current serum used in tissue culture is cattle serum. Human serum, horse serum is used for some specific cells. Cattle serum has several advantages as used in cell culture: adequate resource, mature preparation technique, long application time.

Cattle serum includes bovine calf serum, newborn calf serum and fetal bovine serum. Take sample for cattle serum from Gibco Life Technologies Company, fetal bovine serum derives from caesarean section fetal bovine; newborn calf serum comes from newborn calf born within 24h; bovine calf serum comes from calf with 10 to 30 days. Fetal bovine serum has highest quality because the fetal bovine doesn't expose to outside environment and has lowest antibodies and complement.

2.1.2. Main components of serum and its function

Serum is made from plasma by removing hemaleucin and contains various plasma protein, polypeptide, fat, carbohydrate, growth factor, hormones and inorganic mineral, etc. All these substances keep the physiological balance of promoting or inhibiting cell growth. The following table shows the main components of serum and their mean concentration.

Component	Mean concentration	Component	Mean concentration
Na^+	137mol/L	Alkaline phosphomonoesterase	225U/L
K^+	11 mol/L	Lactic dehydrogenase	860U/L
Cl^-	103 mol/L	Insulin	0.4µg/L
SeO_3^{2-}	26µg/L	Thyroid stimulator	1.2µg/L
Ca^{2+}	136mg/L	Folliclestimulating hormone	9.5µg/L
Fibonectin	35 mg/L	Bovine somatotropin	39µg/L
Urea acid	29 mg/L	Prolactin	17µg/L
Creatine	31mg/L	T_3	1.2µg/L
Hemoglobin	113 mg/L	Cholesterol	310µg/L
Bilirubin(total)	4 mg/L	Cortisone	0.5µg/L
Inorganic phosphorus	100mg/L	Testosterone	0.4µg/L
Glucose	1250mg/L	Progesterone	80µg/L
Urea	160mg/L	Prostaglandin E	6µg/L
Total protein	38g/L	Prostaglandin F	12µg/L
Albumin	23g/L	Vitamin A	90µg/L
α2- macroglobulin	3g/L	Vitamin E	1 mg/L
Endotoxin	0.35µg/L	$Fe^{2+}, Zn^{2+}, Cu^{2+}, Mn^{2+}, Co^{2+}, Co^{3-}$,etc	µg/L to ng/L

Table 1. The main components of serum and their mean concentration.

The main function of serum is listed as follow:

a. Provide essential nutrients

Serum contains various amino acids, vitamins, inorganic minerals, fat, and nucleic acid derivatives, which are essential nutrients for cell growth.

b. Provide adherence and extension factor

Many cells cultured *in vitro* have to attach the culture vessel to grow, which is dependent on extracellular matrix. Cells can secrete extracellular matrix *in vivo*, but this ability will decrease or even disappear according to the increment of passages. Serum contains some components, (fibronection, laminin, etc), which can promote cell adherence.

c. Provide hormone and various growth factors

Serum contains various hormones, such as insulin, adrenocortical hormone (hydrocortisone, dexamethasone), steroid hormone (estradiol, testosterone, and progesterone), etc. The growth factors include fibroblast growth factor (FGF), epidermal growth factor (EGF), pleteletdericed growth factor (PDGF), and others.

d. Provide binding protein(s)

The binding proteins carry low molecular weight material. For example, the albumin carries vitamins, fat (fatty acid, cholesterol) and hormones. Transferrin carries iron.

e. Provide protection for some specific cells

Some cells (such as epithelial cells, myeloid cells) can release protease, which can be neutralized by the anti-protease ingredient in the serum. Serum is widely used to terminate the effect of the trypsin. Serum albumin facilitates the serum viscosity and protects the cell from mechanical damage, especially in the suspension cell culture. The   trace elements and ions, such as SeO_3 and Selenium, play very important role in metabolic detoxification

2.1.3. Disadvantages of using serum in tissue culture

The composition of serum is complicated, including favorable components and unavoidable harmful ingredient. The disadvantages of using serum in tissue culture are listed as follow.

a. For most cells, serum is not physiological fluid in vivo. Cells only contact serum during the injury healing or blood coagulation. Thus the utilization of serum may change the normal condition of some cell. Serum may promote the growth of some cell (such as fibroblast) and inhibit the proliferation of other cells (such as epidermal kerotinocyte)

b. Some components may be toxic to cells. Take polyamine oxidase for example, it can react with the polyamine (such as spermine and spermidine) to form the toxic poly-spermine in highly proliferated cells. Furthermore, complements, antibodies and bacteriotoxin can affect the cell growth, or even lead to cell death.

c. Each batch of serum varies from others and the component can not maintain uniformity.

d. The productive process may infect with mycoplasma or virus, which potentially affect cells and lead to the fail experiment and unreliable experimental results.

2.2. Rat tail collagen

Rat tail collagen, used as either a thin layer on tissue-culture surfaces to enhance cell attachment and proliferation, or as a gel to promote expression of cell-specific morphology and function. This product is ideal for coating of surfaces, providing preparation of thin layers for culturing cells, or use as a solid gel. Rat Tail Collagen is suitable for applications using a variety of cell lines including hepatocytes, fibroblasts and epithelial cells

3. Synthetic medium

The synthetic medium is artificial designed and prepared medium. Nowadays the synthetic medium already becomes standardized commodity with wide varieties and convenience to use.

3.1. Basic medium

3.1.1. Basic components

After experimental selection, the simplest medium is minimum essential media (MEM), which contains more than 20 kinds of substance and can be divided to 4 subgroups.

Inorganic salt: $CaCl_2$, KCl, $MgSO_4$, NaCl, $NaHCO_3$, NaH_2PO_4

Amino acid: arginine, cystine, isoleucine, leucine, lysine, methionine, phenylalanine, threonine, tryptophan, histidine, tyrosine and valine.

Vitamins: partial polyoxometalates calcium, choline chloride, folic acid, inositol, nicotinamide, pyridoxine, riboflavin, thiamine

Carbohydrate: glucose

Besides the above substance related to cell growth, the medium usually also uses phenol red as pH indicator.

MEM is not common basic medium and common medium has more than 30 kinds of component, such as RPMI1640, DMEM. These mediums generally add some non-essential amino acids and vitamins, including serine, proline, biotin, Vitamin B12, etc.

3.2. Category

Nowadays, there are more than tens of basic medium available. The most common ones are listed as follow:

a.   MEM

MEM, also called Eagle's minimal essential medium, is a cell culture medium developed by Harry Eagle that can be used to maintain cells in tissue culture. It only contains 12 kinds of non-essential animo acids, glutamine, 8 vitamins and some basic Inorganic salts

b.   DMEM

A variation of MEM, called Dulbecco's modified Eagle's medium (DMEM), (Dulbecco/Vogt modified Eagle's minimal essential medium), contains approximately four times as much of the vitamins and amino acids present in the original formula and two to four times as much glucose. Additionally, it contains iron and phenol red. DMEM is further divide into high-glucose type (4500g/L glucose) and low-glucose type (1000g/L glucose). High-glucose DMEM is suitable for some tumor cells with faster growth speed and difficult attachment, as it is beneficial to retain and grow in one place.

c.   IMDM

It is modified by Iscove basic DMEM and contains 42 ingredients. It includes selenium as well as additional amino acids and vitamins. In addition, this unique medium lacks iron, with potassium nitrate replacing ferric nitrate. It is well suited for difficult proliferating, low-density cell cultures, including hybrid cell selection after cell fusion, selection of transformed cell after DNA transfection.

d.   RPMI1640

Roswell Park Memorial Institute medium, commonly referred to as RPMI, is a form of medium used in cell culture and tissue culture. The initial formula is suitable for growth of the suspension cells, mainly for lymphoid cells. This medium contains a great deal of phosphate and is formulated for use in a 5% carbon dioxide atmosphere. RPMI1640, the most mature improved medium, is suitable for most types of cells, including tumor cells, normal cells, primary culture cells, passage cell. RPMI1640 is one of the most common used medium.

e.   199/109 medium

199 medium is developed by Morgan and his coworkers in 1950 and is one of the earliest culture medium. It was originally developed as a completely defined media formulation for chick embryo cell culture. 199 medium has more than 60 components and contain almost all the amino acids, vitamins, growth hormone, nucleic acid derivative, etc. 109 medium is improved based on 199 medium and better formulated for the cell culture in a serum-free environment.

f.   HamF10/HamF12

Ham's F-10 medium is a classical media designed by Ham to support the growth of mouse and human diploid cells in 1962. Ham's F-12, as improved products, has been used for the growth of primary rat hepatocytes and rat prostate epithelial cells. A clonal toxicity assay using CHO cells has also been reported with Ham's F-12 as the medium of choice. Ham's F-12 is also available with 25 mM HEPES buffer that provides more effective buffering in the optimum pH range of 7.2-7.4

g. McCoy's 5A

In 1959, McCoy and his coworkers reported the amino acid requirements for in vitro cultivation of Novikoff Hepatoma Cells. These studies were performed using Basal Medium 5A, and subsequently modified to create a new medium known as McCoy's 5A Medium. This media has been employed to support the growth of primary cultures derived from adrenal glands, bone marrow (normal), gingiva, lung, mouse kidney, omentum, skin, spleen and other tissues. It is a general purpose medium for primary and established cell lines.

All these medium are already commercialized and one medium have different forms, such as powder or liquid, large pack or pouch pack. The liquid form is subdivided 10x concentrated solution, 2x concentrated solution and working solution. Some medium don't have phenol red and some don't have calcium ion and magnesium. The user can choose product according to the experiment requirement. The basic components of partial medium are listed as follow:

Inorganic salts

Components	MEM	DMEM	IMDM	RPMI 1640	F10	F12	McCoys5A	199
$CaCl_2$	200.00	200.00	165.00	-	33.30	33.20	100.00	200.00
KCl	400.00	400.00	330.00	400.00	285.00	223.60	400.00	400.00
$MgSO_4$	98.00	97.67	98.00	48.84	74.60	-	98.00	98.00
NaCl	6800.00	6400.00	4500.00	6000.00	7400.00	7599.00	5100.00	6800.00
$NaHCO_3$	2200.00	3700.00	3024.00	2000.00	1200.00	1176.00	2200.00	2200.00
NaH_2PO_4	140.00	125.00	125.00	-	-	-	580.00	140.00
KNO_3	-	-	0.076	-	-	-	-	-
$NaSeO_3$	-	-	0.017	-	-	-	-	-
$Ca(NO_3)_2$	-	-	-	100.00	-	-	-	-
$CuSO_4$	-	-	-	800.00	153.70	142.00	-	-
Na_2HPO_4	-	-	-	-	0.03	0.86	-	-
$MgCl_2$	-	-	-	-	-	57.22	-	-
$Fe(NO_3)_3$	-	0.10	-	-	-	-	-	-
$CuSO_4$	-	-	-	-	0.0025	0.0025	-	-
$FeSO_4$	-	-	-	-	0.083	0.083	-	-
KH_2PO_4	-	-	-	-	83.00	-	-	-

Amino acids

Components	MEM	DMEM	IMDM	RPMI 1640	F10	F12	McCoys5A	199
L-ArginineHCl	126.00	84.00	84.00	200.00	211.00	211.00	42.10	70.00
L-Cystine2HCl	31.00	63.00	91.20	65.00	-	-	-	26.00
L-CystineHCl H₂O	-	-	-	-	25.00	35.00	31.50	0.10
L-HistidineHCl H₂O	42.00	42.00	42.00	15.00	23.00	21.00	21.00	22.00
L-Isoleucine	52.00	105.00	105.00	50.00	2.60	4.00	39.40	40.00
L-Leucine	52.00	105.00	105.00	50.00	13.00	13.00	39.40	60.00
L-LysineHCl	73.00	146.00	146.00	40.00	29.00	36.50	36.50	70.00
L-Methionine	15.00	30.00	30.00	15.00	4.50	4.50	15.00	15.00
L-Phenylalanine	32.00	66.00	66.00	15.00	5.00	5.00	16.50	25.00
L-Threomine	48.00	95.00	95.00	20.00	3.60	12.00	17.90	30.00
L-Tryptophan	10.00	16.00	16.00	5.00	0.60	2.00	3.10	10.00
L-Tyrosine2Na2H₂O	52.00	104.00	104.00	29.00	2.62	7.80	26.20	58.00
L-Valine	46.00	94.00	94.00	20.00	3.50	11.70	17.60	25.00
L-Alanine	-	-	25.00	-	9.00	8.90	13.90	25.00
L-Asparagine	-	-	25.00	50.00	15.00	15.00	45.00	-
L-Aspartic acid	-	-	30.00	20.00	13.00	13.00	20.00	30.00
L-Glutamic acid	-	-	75.00	20.00	14.70	14.70	22.10	75.00
L-Glutamine	-	584.00	284.00	300.00	146.00	146.00	219.20	100.00
Glycine	-	30.00	30.00	10.00	7.50	7.50	7.50	50.00
L-Proline	-	-	40.00	20.00	11.50	34.50	17.30	40.00
L-Serine	-	42.00	42.00	30.00	10.50	10.50	26.30	25.00
L-Hydroxyproline	-	-	-	20	-	-	19.70	10.00

Vitamins

Components	MEM	DMEM	IMDM	RPMI 1640	F10	F12	McCoys5A	199
K-Ca-Pantothenate	1.00	4.00	4.00	0.25	0.70	0.50	0.20	0.01
Choline Chloride	1.00	4.00	4.00	3.00	0.70	14.00	5.00	0.50
Folic acid	1.00	4.00	4.00	1.00	1.30	1.30	10.00	0.01
i-Inositol	2.00	7.20	7.20	35.00	0.50	18.00	36.00	0.05
Niacinamide	1.00	4.00	4.00	1.00	0.60	0.04	0.50	0.025
Pyridoxal HCl	1.00	-	4.00	-	-	-	0.50	0.025
Pyridoxine HCl	-	4.00	-	1.00	0.20	0.06	0.50	0.025
Riboflavin	0.10	0.40	0.40	0.20	0.40	0.04	0.20	0.01
Thiamine HCl	1.00	4.00	4.00	1.00	1.00	0.30	0.20	0.01
Biotin	-	-	0.013	0.20	0.024	0.007	0.20	0.01
Bitamin B_{12}	-	-	0.013	0.005	1.40	1.40	2.00	-
Para-aminobenzoic acid	-	-	-	1.00	-	-	1.00	0.05
Niacin	-	-	-	-	-	-	0.50	0.025
Ascorbic acid	-	-	-	-	-	-	0.50	0.05
a-Tocopherol phosphate	-	-	-	-	-	-	-	0.01
Calciferol	-	-	-	-	-	-	-	0.10
Menadione	-	-	-	-	-	-	-	0.01
Vitamin A	-	-	-	-	-	-	-	0.14

Other chemicals:

Components	MEM	DMEM	IMDM	RPMI 1640	F10	F12	McCoys5A	199
D-Glucose	1000.00	4500.00	44500.00	2000.00	1100.00	1802.00	3000.00	1000.00
Phenol red	10.00	15.00	15.00	5.00	1.20	1.20	10.00	20.00
HEPES	-	-	5958.00	-	-	-	5958.00	-
Sodium pyruvate	-	-	110.00	-	110.00	110.00	-	-
Glutathione(reduced)	-	-	-	1.00	-	-	0.50	0.05
Hypoxanthine.Na	-	-	-	-	4.70	4.77	-	0.04
Thymidine	-	-	-	-	0.70	0.70	-	-
Lipoic acid	-	-	-	-	0.20	0.21	-	-
Linoleic acid	-	-	-	-	-	0.08	-	-
Putrescine 2HCl	-	-	-	-	-	0.16	-	-
Bacto-peptone	-	-	-	-	-	-	600.00	-
Thymine	-	-	-	-	-	-	-	0.30

Adenine sulphate	-	-	-	-	-	-	-	10.00
Adenosine-5-triphosphate	-	-	-	-	-	-	-	0.20
Cholesterol	-	-	-	-	-	-	-	0.20
2-deoxy-D-ribose	-	-	-	-	-	-	-	0.50
Adenosine-5-phosphate	-	-	-	-	-	-	-	0.20
Guanine HCl	-	-	-	-	-	-	-	0.30
Ribose	-	-	-	-	-	-	-	0.50
Sodium acetate	-	-	-	-	-	-	-	50.00
Tween 80	-	-	-	-	-	-	-	20.00
Uracil	-	-	-	-	-	-	-	0.30
Xanthine Na	-	-	-	-	-	-	-	0.34

Table 2. The basic components of some medium (mg/L)

3.3. Serum-free medium

Serum free media (SFM) are important tools that allow researchers to grow a specific cell type or perform a specific application in the absence of serum.

3.3.1. Advantages of using serum free media include:

a.   Easier purification and downstream processing
b.   Precise evaluations of cellular function
c.   Increased definition
d.   More consistent performance
e.   Increased growth and/or productivity
f.   Better control over physiological responsiveness
g.   Enhanced detection of cellular mediators

3.3.2. Things to consider in serum-free culture

Overall, cells in serum-free culture are more sensitive to extremes of pH, temperature, osmolality, mechanical forces, and enzyme treatment.

a.   Antibiotics

It is best not to use antibiotics in serum-free media. If you do, we recommend that you use 5- to 10-fold less than you would in a serum- supplemented medium. This is because serum proteins tend to bind a certain amount of the antibiotic added; without these serum proteins the level of antibiotic may be toxic to certain cells.

b.   Higher density

Cells must be in the mid-logarithmic phase of growth with viability >90% prior to adaptation. Sequential adaptation may be necessary.

Seeding cultures at a higher density than normal at each passage during SFM adaptation may help the process. Because some percentage of cells may not survive in the new culture environment, having more cells present will increase the number of viable cells to further passage.

c.   Clumping

Cell clumping often occurs during adaptation to SFM. We recommend that you gently triturate the clumps to break them up when passaging cells.

d.   Morphology

It is not uncommon to see slight changes in cellular morphology during and after adaptation to SFM. As long as doubling times and viability remain good, slight changes in morphology should not be a reason for concern.

3.4. Basic formula of serum-free medium

SFM includes basic culture medium and supplements. Basic culture mediums commonly use HamF12 and DMEM as 1:1 mixture. The supplements include:

a.   Promote adherence substances

Many cells cultured in vitro have to attach the culture vessel for growth and SFM need to add some supplements for promoting attachment and extension, mainly extracellular matrix (fibronection, laminin, etc).

b.   Somatomedin and hormones

Different growth factors need to add different cells. For example, EGF for the Keratinocytes, NGF for the Neurocyte, ECGF for the endothelial cell. The following table summarizes the utilization of some growth factor.

c.   Enzyme inhibitor

Adherent cell need to trypsinize and passage. The SFM must have enzyme inhibitor to stop the activity of trypsin and protect the cell. The commonest enzyme inhibitor is soybean trypsin inhibitor

d.   Binding protein(s) and translocator

The common binding protein and translocators are transferring and bovine serum albumin.

e.   trace element

Selenium is most common trace element.

4. How to use

Sequential adaptation is preferred method for adapting cells to serum-free media (SFM).Because the change from 75% to 100% SFM may be too stressful for your cells, you may need to carry the cells for 2–3 passages in a 10% serum-supplemented medium: 90% SFM mixture. Most cell lines can be considered fully adapted after 3 passages in 100% SFM. Occasionally you may have trouble getting your cells past a certain step even before going 100% SFM. If this happens, go back and passage the cells 2–3 times in the previous ratio of serum-supplemented media to serum-free media.

Growth factor	Target cell(common)	Target cell(specific)	Recommended concentration
EGF	epiblast, mesoblastema	Keratinocytes, fibroblast, chondrocyte, etc	1-20ng/ml
bFGF	Mesoblastema, neuroectoderm	endothelial cell, fibroblast, chondrocyte, myoblast,etc	0.5-10ng/ml
FGF	Mesoblastema, neuroectoderm	fibroblast, vascular cells.	1-100 ng/ml
ECGF	endothelial cell	endothelial cell	1-3mg/ml
IGF-1	Most cells		1-10ng/ml
PDGF	mesenchymal cell	fibroblast, myocyte, neurogliocyte	1-50ng/ml
NGF	sensory cell, sympathetoblast	Neurocyte, neurogliocyte	5-100ng/ml
TGF-α	Stimulate mesenchymal cell		0.1-3ng/ml
TGF-β	Inhibit epiblast		

Table 3. The utilization of some growth factor

4.1. Protein free medium(PFM) and chemical defined medium(CDM)

Protein free medium (PFM) is a proprietary serum-free and protein-free growth medium that does not contain any hormones or growth factors. This medium is optimized for the cultivation of Chinese Hamster Ovary (CHO) cells in addition to many derivatives of this parent line. The absence of protein in the medium eliminates any transmission risk of blood borne diseases, resulting in a product that is far safer than the current culture media on the market.

Chemical defined medium (CDM) is a media in which the chemical nature of all the ingredients and their amounts are known.

5. Other medium

5.1. Balanced salt solution (BSS)

Balanced salt solutions can provide an environment that maintains the structural and physiological integrity of cells in vitro. Solutions most commonly include sodium, potassium, calcium, magnesium, and chloride. Balanced salt solutions are used for washing tissues and cells and are usually combined with other agents to treat the tissues and cells. They provide the cells with water and inorganic ions, while maintaining a physiological pH and osmotic pressure. The following table lists the formula of several common BSS.

	Ringer	PBS	Tyrode	Earle	Hank's	D-Hank's	Dulbecco
NaCl	9.00	8.00	8.00	6.80	8.00	8.00	8.00
KCl	0.42	0.20	0.20	0.40	0.40	0.40	0.20
CaCl₂	0.25	-	0.20	0.20	0.14	-	0.10
MgCl₂6H₂O	-	-	0.10	-	-	-	0.10
MgSO₄7H₂O	-	-	-	0.20	0.20	-	-
Na₂HPO₄H₂O	-	1.56	-	-	0.06	0.06	-
NaH₂PO₄2H₂O	-	-	0.05	0.14	-	-	1.42
KH₂PO₄	-	0.20	-	-	0.06	0.06	0.20
NaHCO₃	-	-	1.00	2.20	0.35	0.35	-
Glucose	-	-	1.00	1.00	1.00	-	-
Phenol red	-	-	-	0.02	0.02	0.02	0.02

Table 4. The formula of several common BSS

5.2. Medium used for digestion

Primary cell cultures usually need to digest and dissociate the tissue to cell suspension. Continuous passage cultures need to digest the adherent cell from culture dishes. The common digestion solutions are trypsin, EDTA solution and collagenase solution.

5.2.1. Trypsin

In a tissue culture lab, trypsins are used to re-suspend cells adherent to the cell culture dish wall during the process of harvesting cells. Trypsin is used to cleave proteins bonding the cultured cells to the dish, so that the cells can be suspended in fresh solution and transferred to fresh dishes.

Trypsin is an endopeptidase produced by the gastro-intestines of mammals, and has an optimal operating pH of about 8 and an optimal operating temperature of about 37 °C. The biochemical assays performed on Trypsin 1:250 determine both trypsin specific-activity at the level of certain co-purified enzymes that influence cell removal and viability

5.2.2. EDTA solution

In Cell Culture applications, EDTA is used for its chelating properties which binds to calcium and other ions and thus prevents adjoining of cadherins(i.e. the integral membrane proteins involved in calcium-dependent cell-adhesion) between cells, preventing the clumping of cells growing in liquid suspension or even detaching adherent cells for passaging. The working concentration is 0.02%. Trypsin//EDTA is a combined method for detaching cells.

5.2.3. Collagenase solution

Collagenase is especially valuable when tissues are too fibrous or too sensitive to allow the use of trypsin, which is ineffective on fibrous material and damaging to sensitive material. Dissociation is usually achieved either by perfusing whole organs or by incubating smaller pieces of tissue with enzyme solution. The working solution is 0.1-0.3mg/ml or 200000U/L. The optimal operating pH is about 6.5.

5.3. pH adjusting medium

5.3.1. NaHCO$_3$ solution

Sodium bicarbonate is used as part of a buffering system commonly used to maintain physiological pH 7.2 - 7.4 in a culture environment. The concentration of the sodium bicarbonate in the medium must be matched with the level of CO_2 in the atmosphere above the medium. For media containing 1.5 to 2.2 g/L sodium bicarbonate, use 5% CO_2. For media containing 3.7 g/L sodium bicarbonate, use 10% CO_2.

5.3.2. HEPES

HEPES (4-(2-hydroxyethyl)-1-piperazineethanesulfonic acid) is a zwitterionic organic chemical buffering agent. The level of HEPES in cell culture media may vary from 10mM to 25mM. It is added to the media solely for extra buffering capacity when cell culture requires extended periods of manipulation outside of a CO_2 incubator.

5.4. Antibiotics

Researchers use antibiotics and antimycotics to prevent contamination. The common ones are Penicillin & Streptomycin (P/S). Penicillin Streptomycin mixtures contain 5,000 units of penicillin (base) and 5,000 µg of streptomycin (base)/ml utilizing penicillin G (sodium salt)

and streptomycin sulfate in 0.85% saline. The following table summarizes the utilization of different antibiotics.

Antibiotic	Storage Conditions	Working Concentration	Stability in Culture (37°C)
Amphotericin B	-20°C	2.5 µg/ml	~3 days
Ampicillin	4°C	100 µg/ml	~3 days
Chloramphenicol	-20°C	5 µg/ml	~5 days
Ciprofoxacin	-20°C	10 µg/ml	not known
Gentamicin	4°C	50 µg/ml	~5 days
Hygromycin B	4°C	500 µg/ml	not known
Kanamycin	-20°C	100 µg/ml	~5 days
Neomycin	-20°C	50 µg/ml	~5 days
Penicillin	-20°C	100 U/ml	~3 days
Puromycin	-20°C	20 µg/ml	not known
Streptomycin	-20°C	100 µg/ml	~3 days
Tetracycline	-20°C	10 µg/ml	~4 days

Table 5. Antibiotics Storage Conditions Working Concentration Stability in Culture (37°C)

5.5. Glutamine

Glutamine is an unstable essential amino acid required in cell culture media formulations. Most commercially available media are formulated with free L-glutamine which is either included in the basal formula or added to liquid formulations at time of use. The concentration of L-glutamine used in classical media is usually 0.002mol/L.

Author details

Zhanqiu Yang and Hai-Rong Xiong
School of Basic Medical Sciences, Wuhan University, The People's Republic of China

6. References

Barnes, D.W. (1987). Serum-free animal cell culture. BioTechniques 5:534-541.

Barnes, D.W., Sirbasku, D.A., and Sato, G.H. (eds.) (1984). Cell Culture Methods for Cell Biology,Vols. 1-4. Alan R. Liss, New York.

Davis, J.M. (ed.) (1994. Basic Cell Culture: A Practical Approach. Oxford University Press, Oxford.

Eagle, H. 1955. Nutrition needs of mammalian cells in tissue culture. Science 122:501-504.

Eagle, H. 1973. The effect of environmental pH on the growth of normal and malignant cells. Journal of Cellular Physiology. 82:1-8.

Ham, R.G. 1984. Formulation of basal nutrient media.In Cell Culture Methods for Cell Biology, Vol. 1 (D.W. Barnes, D.A. Sirbasku, and G.H.Sato, eds.) pp. 3-21. Alan R. Liss, New York.

Sato, J.D., Hayashi, I., Hayashi, J., Hoshi, H.,Kawamoto, T., McKeehan, W.L., Matsuda, R., Matsuzaki, K., Mills, K.H.G., Okamoto, T., Serrero, G., Sussman, D.J., and Kan, M. 1994. Specific cell types and their requirements. In Basic Cell Culture: A Practical Approach (J.M. Davis, ed.) pp. 181-222. Oxford University Press, Oxford.

Chapter 2

Contamination of Tissue Cultures by Mycoplasmas

Shlomo Rottem, Nechama S. Kosower and Jonathan D. Kornspan

Additional information is available at the end of the chapter

1. Introduction

Mycoplasmas are the smallest and simplest self-replicating bacteria [1]. These microorganisms lack a peptidoglycan based rigid cell wall and thus are not susceptible to antibiotics, such as penicillin and its analogues, which are effective against most bacterial contaminants of cell cultures. The trivial name mycoplasma encompasses all species included in the class *Mollicutes*: i.e. the genera *Mycoplasma, Acholeplasma, Spiroplasma, Anaeroplasma and Ureaplasma*. Because mycoplasmas have an extremely small genome (0.58–2.20 Mb compared with the 4.64 Mb of *Escherichia coli*), these organisms have limited metabolic options for replication and survival. The smallest genome of a self-replicating organism known at present is the genome of *Mycoplasma genitalium* (0.58 Mb; Ref. 2). Comparative genomic studies suggested that the genome of this organism still carries almost double the number of genes included in the minimal gene set essential for cellular function [3]. Owing to their limited biosynthetic capabilities, most mycoplasmas are parasites, exhibiting strict host and tissue specificities [4]. The aim of this review is to collate present knowledge on the strategies employed by mycoplasmas while interacting with tissue culture cells. Prominent among these strategies is the adherence of mycoplasmas to host cells, the invasion of mycoplasmas into host cells and the fusion of mycoplasmas with host cells. We shall discuss the intriguing questions of how a mycoplasma infecting tissue culture cells subvert and damage the host cells by mediating transformation of the cells, affecting the signal-transduction pathways and the metabolism of immune and non-immune cells. We shall also present and discuss the common procedures for isolation, identification and eradication of a mycoplasma contamination of tissue cultures.

2. Mycoplasmas contaminating cultured cells

It is well established that stable cell cultures are frequently contaminated by mycoplasmas. In a study carried out in the USA at the Food and Drug Administration (FDA), over 20,000 cell cultures were examined during a period of 30 years, 15% of which were found to be contaminated [5]. Higher incidences of contamination have also been reported. Three different surveys in Japan showed an incidence of mycoplasma contamination of 60-80%, and an incidence of 65% was reported in Argentina [5]. At least 20 distinct *Mycoplasma* or *Acholeplasma* species have been isolated from contaminated cell lines. Ninety-five percent of the contaminants were identified as either *M. orale, M. arginini, M. hyorinis, M. fermentans* or *A. laidlawii* [5]. although the frequency of isolation of a particular species varies with the particular study.

In general, primary cell cultures are less frequently contaminated than continuous cell lines. However, since many viral vaccines (such as those for measles, mumps, rubella, polio and rabies) are produced in primary cell cultures, many countries require such cultures to be screened carefully for mycoplasma contamination before approval can be given for release of the vaccine (or other biological products intended for human use) to the market-place.

All cell types, including virus-infected, transformed, or neoplastic cell cultures grown in monolayers and/or in suspension, derived from all host-types examined, are subject to contamination. Mammalian and avian cell lines were the most commonly contaminated although, on occasions, cell cultures derived from reptiles, fish, insects or plants were also contaminated. Most studies have examined fibroblast cell cultures, but epithelial, endothelial, lymphocytic and hybridoma cell-culture lines have also been found to be contaminated. Frequently, the number of mycoplasmas far exceeds (often by 1000-fold) the number of tissue-culture cells in an infected cell culture. The information available on the contamination of cultures of differentiated cell lines is limited, and more data are needed before a proper assessment can be made. Mycoplasma contamination of vaccines presents a potential health hazard; consequently, identifying the source(s) of contamination is a key concern. The probable source of most mycoplasma contaminants in primary cell culture is the original tissue used to develop the primary cell culture lot. Whereas lung, kidney, or liver tend to be mycoplasma-free, the foreskin, the lower female-urogenital tract, or tumor tissues, are subject to mycoplasma colonization, and generally show a higher rate of contamination [5]. Nonetheless, contamination from exogenous sources also occurs during cell propagation and continuous cell cultures are the most frequently contaminated. The main source of contamination is, in many cases, infection by previously-contaminated cell cultures that have been maintained and processed in the same laboratory [5]. Mycoplasmas are spread by using laboratory equipment, media, or reagents that have been contaminated by previous use in processing mycoplasma-infected cell cultures. New cell-culture acquisitions should be quarantined, tested and guaranteed mycoplasma-free before introduction into the tissue-culture laboratory. Common experimental stock materials, such as virus pools, or monoclonal antibody preparations, can also be a key source of mycoplasma contamination. As there is no legal requirement for suppliers to provide mycoplasma-free products, bovine serum should be considered as a possible source of contamination. Mycoplasma contaminants of bovine serum are primarily bovine species, with *A. laidlawii* and *M. arginini* being isolated most frequently [5].

3. Mode of interaction with host cells

3.1. Adherence to host cells

Most mycoplasmas are typical extracellular microorganisms able to adhere to the surface of tissue culture cells. Many mycoplasmas exhibit the typical polymorphism of mycoplasmas, with the most common filamentous, flask shapes or ovoid structures (Figure 1, Ref. 6). The adherence of mycoplasmas to host cells is an initial and essential step in tissue colonization [4]. The lack of a cell wall has forced mycoplasmas to develop sophisticated molecular mechanisms to enable their prolonged adhesion. Adherence is associated with adhesins as well as host cell receptors that mediate interaction of the bacteria with the host cells [7].

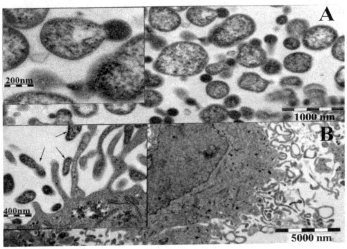

Figure 1. Transmission electron microscopy of *M. hyorhinis* (A) and of a melanoma cell culture infected by *M. hyorhinis* (B). Flask shaped bacteria in close proximity to melanoma cells are indicated by arrows.

A polar, tapered cell extension at one of the poles containing an electron-dense core in the cytoplasma was described in some mycoplasmas (Figure 2). This structure, termed the tip organelle, functions mainly as an attachment and motility organelle. A variety of surface proteins that participate in the adhesion process are densely clustered at the tip organelle [4]. The role of host cell surface sialoglycoconjugates as receptors for mycoplasmas has been suggested [8]. The carbohydrate moiety of the glycoprotein, which serves as a receptor for *M. pneumoniae* on human erythrocytes, has been identified as having a terminal NeuAc(α2–3)Gal(β1–4)GlcNAc sequence [9]. Nevertheless, neuraminidase treatment has frequently failed to abolish the ability of various eukaryotic cells to bind *M. pneumoniae* [10]. A sialic acid-free glycoprotein, isolated from cultured human lung fibroblasts, which serves as a receptor for *M. pneumoniae*, has been isolated by Geary et al. [11]. Sulfated glycolipids containing terminal Gal(3SO$_4$)β1 residues were also found to function as receptors [12]. Clearly, there is more than one type of receptor for the various adhering mycoplasmas.

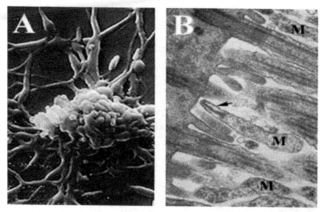

Figure 2. A, scanning electron microscopy of filamentous *M. pneumoniae*. B, transmission electron microscopy of flask-shaped *M. pneumoniae* (M) attached by the terminal tip organelle (arrow) to ciliated mucosal cells. Magnification: A, x10,000; B, x36,000.

The attachment of mycoplasmas to the surface of host cells may interfere with membrane receptors or alter transport mechanisms of the host cell. The disruption of the K^+ channels of ciliated bronchial epithelial cells by *M. hyopneumoniae* that resulted in ciliostasis has been described [13]. The host cell membrane is also vulnerable to toxic materials released by the adhering mycoplasmas. Although toxins have not been associated with mycoplasmas, the production of cytotoxic metabolites and the activity of cytolytic enzymes are well established. Oxidative damage to the host cell membrane by peroxide and superoxide radicals excreted by the adhering mycoplasmas appears to be experimentally well-substantiated [14]. The intimate contact of the mycoplasma with the host cell membrane may also result in the hydrolysis of host cell phospholipids catalyzed by the potent membrane-bound phospholipases present in many mycoplasma species [15]. This could trigger specific signal cascades [16] or release cytolytic lysophospholipids capable of disrupting the integrity of the host cell membrane [17, 18].

3.2. Invasion of host cells

It is generally accepted that mycoplasmas remain attached to the surface of host cells [1]. However, some mycoplasmas have evolved mechanisms for entering host cells that are not naturally phagocytic. The intracellular location is obviously a privileged niche, well protected from the action of many antibiotics. Mycoplasma invasion of host cells was intensively studied with *M. penetrans*, isolated from the urogenital tract of acquired immunodeficiency syndrome (AIDS) patients [19, 20]. It was shown that this microorganism has invasive properties and localizes in the cytoplasm and perinuclear regions [21, 22, 23]. Mycoplasmal invasion of host cells is a complex process that involves a variety of mycoplasmal and host cell factors. It is likely that surface molecules (proteins and lipids) that facilitate the adhesion process of mycoplasmas will have an effect on the invasion. Nevertheless, adherence to the surface of host cells is not sufficient to trigger events that lead to invasion. The signals generated by the interaction of host cells with invasive mycoplasmas have yet to be investigated. It has been shown that bacterial invasion is based on the ability of several bacteria to bind sulfated polysaccharides or fibronectin [24]. It was suggested that these compounds form a molecular bridge between the bacteria and eukaryotic surface proteins [25] that enables invasion. Fibronectin binding activity was detected in *M. penetrans*. This organism, which contains a 65-kDa fibronectin binding protein, binds selectively immobilized fibronectin [23]. The finding that *M. fermentans* binds plasminogen (Plg) and in the presence of urokinase-type Plg activated (uPA) internalization was apparent (26, 27), indicates that the ability of *M. fermentans* to invade host cell stems from its potential to bind and activate Plg to plasmin, a protease with broad substrate specificity. Plg and uPA are two proteins that play an important role in the invasion of several human malignant tumors [28], therefore it is not surprising that the same system stimulates *M. fermentans* invasion. Other mycoplasmas known to be surface parasites such as *M. pneumoniae* [29], *M. genitalium* [30], *M. gallisepticum* [31], and *M. hyorhinis* [6] were also found under certain circumstances to reside within host cells.

In studying bacterial invasion, it is essential to differentiate between microorganisms adhering to a host cell and those which have penetrated the cell. The light microscopic and electron microscopic observations of mycoplasmas engulfed in membrane vesicles lead to conflicting interpretations. It is not clear whether mycoplasmas are intra, or are they at the bottom of crypts formed by the invagination of the cell membrane [32]. A more sophisticated ultrastructural study was based on a combined immunochemistry and electron microscopy approach. Staining surface polysaccharides of the host cell with ruthenium red allows a better differentiation between intracellular and extracellular mycoplasmas [33]. Currently, the gentamicin resistance assay is the most common assay to differentiate intracellular from extracellular bacteria [7, 34]. In this assay, the extracellular bacteria are killed by gentamicin, but the intracellular bacteria are shielded from the antibiotic because of the limited penetration of the gentamicin into eukaryotic cells. The gentamicin procedure was successfully adapted to mycoplasma systems [21, 31]. Usually the number of intracellular bacteria is determined by washing the host cells free of the antibiotic, lysing them with mild detergents to release the bacteria and counting the colonies [35]. Because mycoplasmas are as susceptible to detergent lysis as the host cells, dilutions of the mycoplasma-infected host cells should be plated directly onto solid mycoplasma media without lysing them beforehand. Each mycoplasma colony represents one infected host cell rather than a single intracellular mycoplasma [34].

Immunofluorescent staining of internalized bacteria and of those remaining on the cell surface, combined with confocal laser scanning microscopy, has demonstrated that several mycoplasmas penetrate eukaryotic cells (Figure 3; Refs. 22, 36) This nondestructive, high-resolution method allowed infected host cells to be optically sectioned after fixation and immunofluorescent labeling. Imaging single infected HeLa cells revealed that invasion is both time and temperature dependent. Penetration of melanoma cells by *M. hyorhinis* has been observed as early as 30 min after infection [6], whereas invasion of cultured HEp-2 cells by *M. penetrans* has been shown to begin after 2 h of infection [36].

The intracellular fate of invading bacteria can vary greatly. Most invasive bacteria appear to be able to survive intracellularly for extended periods of time, at least if they have reached a suitable host cell [37]. Other engulfed bacteria are degraded intracellularly via phagosome-lysosome fusion. The invasive bacteria either remain and multiply within the endosomes after invasion or are released via exocytosis and/or the lysis of the endosomes which may allow multiplication within the cytoplasm. Most ultrastructural studies performed with engulfed mycoplasmas revealed mycoplasmas within membrane-bound vesicles [30, 33, 38]. Persistence of *M. penetrans* within NIH/3T3 cells, Vero cells, human endothelial cells, HeLa cells, WI-38 cells, and HEp-2 cells has been observed over a 48–96 h postinfection [19, 23]. *M. gallisepticum* remains viable within HeLa cells during 24–48 h of intracellular residence [31]. The observation of vesicles stuffed with *M. penetrans* in various host cells was taken as an indication that *M. penetrans* is able to divide within intracellular vesicles of the host cells [19]. Nonetheless, the intracellular multiplication of mycoplasmas remains to be convincingly demonstrated.

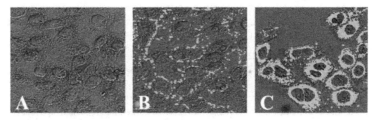

Figure 3. Confocal micrographs demonstrating binding and internalization of *M. hyorhinis* (green fluorescence) by melanoma cells. A, Control of uninfected melanoma cells; B, Formaldehyde fixed melanoma cells infected with mycoplasma (bacteria on the melanoma cell surface); C, Native melanoma cells infected by mycoplasma (bacteria internalized by cells).

Almost all invasive bacteria that come into contact with the host cell surface trigger cytoskeletal rearrangements that facilitate bacterial internalization [35, 39]. Involvement of the host cell cytoskeleton in internalization is considered to be the result of a host cell signal transduction cascade induced by the invasive bacterium. As in many signal transduction processes initiated by bacteria, kinases and/or phosphatases are usually involved [39]. The invading mycoplasmas generate uptake signals that trigger the assembly of highly organized cytoskeletal structures in the host cells [23]. Yet, the nature of these signals and the mechanisms used to transduce them are not fully understood. Specific activation of protein kinases occurs during the internalization of most of the bacteria taken up by microtubule-dependent mechanisms [16]. It has been shown that invasion of HeLa cells by *M. penetrans* is associated with tyrosine phosphorylation of a 145-kDa host cell protein [21]. Tyrosine phosphorylation activates phospholipase C to generate two second messengers: phosphatidylinositol metabolites and diacylglycerol (DAG). Changes in host cell lipid turnover occur as a result of *M. penetrans* binding and/or invasion of Molt-3 lymphocytes

[40]. These changes include the accumulation of DAG and the release of unsaturated fatty acids, predominantly long-chain polyunsaturated ones such as docosahexanoic acid ($C_{22:6}$, 40]. Nonetheless, metabolites of phosphatidylinositol were not detected. These observations support the hypothesis that *M. penetrans* stimulates host phospholipases to cleave membrane phospholipids, thereby initiating the signal transduction cascade. Because in HeLa cells, which are invaded by *M. penetrans*, DAG is generated, it is likely that the protein kinase C is activated in the host cells. Indeed, transient protein kinase C activation was demonstrated in invaded HeLa cells by several methods, including translocation to the plasma membrane and enzymatic activity [22]. However, activation was weak and transient, peaking at 20 min postinfection. How any of these different signal transduction events lead to specific microtubule activity resulting in mycoplasmal internalization is unknown. The role of these signals in the penetration, survival, and proliferation of mycoplasmas within host cells, as well as the involvement of the lipid intermediates in the pathobiological alterations taking place in the host cells, merit further investigation.

3.3 Fusion with host cells

The lack of a rigid cell wall allows direct and intimate contact of the mycoplasma membrane with the cytoplasmic membrane of the eukaryotic cell. Under appropriate conditions, such contact may lead to cell fusion. Fusion of mycoplasmas with eukaryotic host cells has been first observed in electron microscopic studies [41]. The development of energy transfer and fluorescence methods has enabled investigation of the fusion process on a quantitative basis in an experimental cell culture-mycoplasma system and has also allowed the identification of fusogenic mycoplasmas. In all the fusogenic *Mycoplasma* species tested, fusogenicity is dependent on the unesterified cholesterol content of the cell membrane [42]. Fusogenic activity can be found only among mycoplasmas requiring unesterified cholesterol for growth, whereas species, which do not require cholesterol, are nonfusogenic. Among the *Mycoplasma* species, the human mycoplasma, *M. fermentans*, is highly fusogenic, capable of fusing with a variety of cells [2]. It is widely accepted that the reorganization of the membrane structure that occurs during fusion requires that the lipid bilayer is broken up and that other inverted configurations, such as reversed nonbilayer aggregates, are being formed [43, 44, 45]. It has been shown that the polar lipid fraction of *M. fermentans* is capable of enhancing the fusion of small, unilamellar phosphatidylcholine-cholesterol (1:1 molar ratio) vesicles with Molt-3 lymphocytes in a dose-dependent manner, suggesting that a lipid component acts as a fusogen [17, 46]. In an attempt to identify the fusogen, detailed lipid analyses of *M. fermentans* membranes were performed [17, 47, 48], revealing that the polar lipid fraction of this organism contains unusual choline-containing ether phosphoglycolipids, 1-O-alkyl/alkenyl-2-O-acyl-glycero-3-phosphocholine and its lyso-form 1-O-alkyl/alkenyl-glycero-3-phosphocholine [49]. The ether lipids, mainly the lyso-derivative has a marked effect on the fusion of *M. fermentans* with host eukaryotic cells [50]. Very little is known about the role of membrane proteins in the fusion process. The observation that fusion of *M. fermentans* with Molt-3 cells was inhibited by pretreatment of intact *M. fermentans* with proteolytic enzymes [51] implies that this organism possesses a proteinase-sensitive receptor(s) responsible for binding and/or the establishment of tight contact with the cell surface of the host cell involved in fusion. During the fusion process, mycoplasma components may be delivered into the host cell and affect the normal functions of the cell. A whole array of hydrolytic enzymes has been identified in mycoplasmas [1, 15, 52]. Most remarkable are the mycoplasmal nucleases [1] that may degrade host cell DNA. It has recently been shown that *M. fermentans* contains a potent phosphoprotein phosphatase [52]. The delivery of an active phosphoprotein phosphatase into the eukaryotic cell upon fusion may interfere with the normal signal transduction cascade of the host cell.

4. Effects of mycoplasmas on cell cultures

Effects on cell function and metabolism have long been recognized as common in mycoplasma contaminated cell cultures. The nature of the effects depends on the contaminating species and strain of mycoplasma, and on the type of cell infected. Frequently, the effects are due to nutrient deprivation, such as the depletion of amino acids, sugars, fatty acids, cholesterol or nucleic-acid precursors [5], the depletion of choline [4] or the activity of mycoplasmal endonucleases [53], mycoplasmal arginine deiminase [54] or mycoplasmal thymidine phosphorylase [55]. Some mycoplasmas have been shown to produce severe cytopathic effects (CPE) characterized by stunted, abnormal growth and rounded, degenerated cells, apparently due to the promotion or inhibition of apoptosis [56]. The promotion of apoptosis may be due to direct effects of mycoplasma components. Thus, *M. bovis* infection sensitizes some host cells to apoptosis through participation of mycoplasmal endonucleases [53]. Choline deficiency induced by *M. fermentans* enhances rat astrocyte apoptosis [4]. Some mycoplasmas promote host cell death via induction of pro-apoptotic genes [57, 58]. Pro-apoptotic and anti-apoptotic mycoplasmas appear to alter apoptosis regulatory genes differently [59].

4.1. Competition for precursors

Genomic analyses of mycoplasmas have revealed the limited biosynthetic capabilities of these microorganisms [60, 61]. Mycoplasmas apparently lost almost all the genes involved in the biosynthesis of amino acids, fatty acids, cofactors, and vitamins and therefore depend on the host microenvironment to supply the full spectrum of biochemical precursors required for the biosynthesis of macromolecules [1]. Competition for these biosynthetic precursors by mycoplasmas may disrupt host cell integrity and alter host cell function. Nonfermenting *Mycoplasma spp.* utilize the arginine dihydrolase pathway for generating ATP [62] and rapidly deplete the host's arginine reserves affecting protein synthesis, growth and host cell divisions. The effect on the cellular genome may be expressed in chromosomal breakage, multiple translocation events, reduction in chromosome number and the appearance of new and/or additional chromosome variants [63]. Since histones are arginine rich, it was suggested that mycoplasmas may exert their effects on cellular genomes by depleting arginine and thus inhibiting histone synthesis [62]. However, as fermenting mycoplasmas also induce chromosomal aberrations, other mechanisms, including competition for nucleic acid precursors, or degradation of host-cell DNA by mycoplasma nucleases, may be involved. *M. fermentans* infection of cell cultures has been shown to result in a choline-deficient environment and in the induction of apoptosis [64]. Choline is an essential dietary component that ensures the structural integrity and signaling functions of the cell membranes; it is the major source of methyl groups in the diet, and it directly affects lipid transport and metabolism and the cholinergic neurotransmission and transmembrane signaling of cells of the nervous system [65].

4.2. Cytopathic effects

Mycoplasmal attachment to eukaryotic cells may sometimes lead to a pronounced cytopathic effect. Attachment permits the mycoplasma contaminant to release noxious enzymatic and cytolytic metabolites directly onto the tissue cell membrane. Some mycoplasmas selectively colonize defined areas of the cell culture. This results in microcolony formation producing microlesions and small foci of necrosis, e.g., *M. pulmonis*, or form plaques, e.g., *M. gallisepticum*, in an agar overlay system [5]. Microcolonization suggests that mycoplasma-specific receptors are localized in defined areas of the cell monolayer. However, other fermenting mycoplasmas, e.g., *M. hyorhinis*, attach to every cell and destroy the entire monolayer, producing a generalized cytopathic effect. With HeLa cells infected by the invasive *M. penetrans*, the most pronounced effect was the vacuolation of the host cells [22]. The vacuoles appeared to be empty, differing from the described membrane-bound vesicles containing clusters of bacteria [19]. The number and size of the vacuoles depended on duration of infection. Because vacuolation is not obtained with *M. penetrans* cell fractions [22], it is unlikely that a necrotizing cytotoxin is involved in the generation of the cellular lesions. A possible mechanism that leads to vacuolation may be associated with the accumulation of organic peroxides upon invasion of HeLa cells by *M. penetrans*. Indeed, when HeLa cells were grown with the antioxidant α-tocopherol, the level of accumulated organic peroxides was extremely low, and vacuolation was almost completely abolished [22].

Being unable to synthesize nucleotides, mycoplasmas developed potent nucleases, either soluble ones secreted into the extracellular medium or membrane-bound nucleases [1, 66, 67] apparently as a means of producing nucleic acid precursors required for metabolism. It has been shown that, occasionally, secreted mycoplasmal nucleases are taken up by the host cells [68]. Thus, it was suggested that the cytotoxicity of *M. penetrans* is mediated at least in part by a secreted mycoplasmal endonuclease that is cleaving DNA and/or RNA of the host cells [66], and the endonuclease activity of *M. bovis* was implicated in the increased sensitivity of lymphocytic cell lines to various inducers of apoptosis [69].

4.3. Transformation of cells mediated by mycoplasmas

Cell culture contamination may go undetected because mycoplasma infections do not produce the overt turbid growth that is commonly associated with bacterial and fungal contamination. Mycoplasma growth can grow in close interaction with mammalian cells, often silently for a long period of time. However, prolonged interactions with mycoplasmas with seemingly low virulence could, through a gradual and progressive course, induce chromosomal instability as well as malignant transformation, promoting tumorous growth of mammalian cells [70, 71]. Mycoplasmal-induced malignant transformation is a multistage process [70] associated with increased or decreased expression of many genes, especially cancer-related genes [72]. Over expression of H-*ras* and c-*myc* oncogenes were found to be closely associated with both the initial reversible and the subsequent irreversible states of the mycoplasma-mediated transformation of cells [71]. In some cases, mycoplasmas have been shown to induce the production of proteins that play essential roles in the development of malignancy. Examples are the mycoplasmal-promoted production in diverse types of cultured cells of bone morphogenetic protein 2 (BMP2) that enhances tumor growth by increasing cell proliferation [73]; mycoplasma-induced diminished activation of the tumor suppression protein p53, and enhanced fibroblast transformation by the oncogenic H-*ras* [74] ; promotion of cancer cell motility and migration by P37, the major immunogen of *M. hyorhinis*, through activation of the matrix metalloproteinase-2 [75].

4.4. Modulation of immune and non-immune cell metabolism

The effects of mycoplasmas on the immune system are well established and include effects on differentiation and activation of innate immunity cells (macrophages, dentritic cells, neutrophils, NK) and on adaptive immunity cells (T and B cells). Mycoplasma and mycoplasmal components are potent macrophage activators, and stimulate the release of various proinflammatory cytokines, such as tumor necrosis factor α (TNFα), interleukin-1(IL-1), IL-6, NO [4, 76]. In turn, some cytokines participate in lymphocyte differentiation and maturation [4]. *M. fermentans* induces a partial differentiation of the human monocytic cell line THP-1 [77]. Mycoplasma-contaminated exosome fractions of dentritic cells are mitogens for naive B lymphocytes and promote immunoglobulin secretion [78].

Mycoplasmas and mycoplasmal components interact with diverse non-immune cells [56, 57, 58, 79], with some information available on the cellular proteins affected by them. *M. salivarium* and *M. fermentans* induce the cell surface expression of intercellular adhesion molecule 1 (ICAM-1) in human gingival fibroblasts [80]. Hyperammonia toxicity in irradiated hepatoma cells has been shown to be due to contamination by mycoplasma containing arginine deiminase, that converts arginine to citrulline and ammonia [54]. *M. pneumoniae* induces the expression of the major airway protein mucin (MUC5AC) in cultured airway epithelial cells isolated from asthmatic subjects, but not in cells isolated from normal subjects; the preferential expression of MUC5AC in cells isolated from asthmatic subjects suggests that asthmatic epithelial cells may be primed to respond to the mycoplasma [81], thus pointing to the importance of identifying consequences of mycoplasma contamination that may be observed only in certain specific types of cultured cells. Catabolic mycoplasmal enzymes may interfere with chemotherapy. This is illustrated by the finding that the antiviral and cytostatic activity of pyrimidine nucleoside analogues (used as chemotherapeutic agents) is markedly decreased in *M. hyorhinis* contaminated cells, due to the mycoplasmal thymidine phosphorylase that degrades pyrimidine nucleoside analogues [55]. Contamination of human cultured neuroblastoma SH-SY5Y and melanoma cell lines by *M. hyorhinis* results in increased levels of calpastatin (the endogenous inhibitor of the ubiquitous Ca^{2+}-dependent protease calpain). The calpastatin upregulation resides in the *M. hyorhinis* lipoprotein fraction (LPP), via the IκB/NF-κB transcription pathway [79]. LPPs of several other mycoplasma species have also been found to upregulate calpastatin [J.D. Kornspan, T. vaisid, S. Rottem and N.S. Kosower, unpublished data]. Amyloid-β-peptide and Ca^{2+} (these are central to the pathogenesis of Alzheimer's Disease) activate calpain and are toxic to neuroblastoma cultured cells. The increased calpastatin levels in the mycoplasma-infected cells attenuate the calpain-related amyloid-β-peptide and Ca^{2+}-toxicity. Calpain and calpastatin are widely distributed in biological systems, with the ratio of calpastatin to calpain varying among cells. The calpain-calpastatin system has been implicated in a variety of cellular physiological and pathological processes [82]. Since calpastatin level is important in the control of calpain activity, mycoplasmas may play a role in a variety of metabolic and signal transduction pathways in some types of cultured cells. The mycoplasma-induced elevation of calpastatin provides an example of mycoplasmal effects on intracellular proteins in non-immune cells, resulting in important alterations in the host cell functions.

4.5. Effect on virus infection

Mycoplasmas may alter the progress of viral infections in cell cultures [83, 84]. As mycoplasmas may also cause virus-like CPE, many investigators have mistaken cytolytic mycoplasmas for viruses. Like viruses, mycoplasmas are filterable, hemadsorbant, hemagglutinant, resistant to certain antibiotics, able to induce chromosomal aberrations, and sensitive to detergents, ether and chloroform; thus the first established mycoplasma pathogens of humans (*M. pneumoniae*), animals (*M. mycoides*) or plants (*Spiroplasma* spp.) were believed to be viruses. Some mycoplasmas have no detectable effect on viral growth. Others can decrease, or even increase, virus yields in infected cell culture [85]. The effect depends on the strain or species of mycoplasma, the virus, and the cell culture used. At least two mechanisms responsible for decreasing viral yields in vitro have been identified. The cytolytic, fermenting mycoplasmas suppress metabolism and growth, resulting in a decrease in viral yields. Arginine-utilizing mycoplasmas decrease the titers of arginine-requiring DNA viruses by depleting arginine from the medium [62]. Mycoplasmas may render cell cultures less sensitive to exogenously supplied interferon and thus to increase virus yields [86]. Mycoplasmas may also inhibit viral transformation of cell cultures by known oncogenic viruses [5, 87].

4.6. Signal transduction pathways

Mycoplasmas and mycoplasmal membrane LPPs attach to certain Toll-like receptors (TLRs) of the host cell membrane. The main TLR involved appears to be TLR2, with participation of TLR6 as coreceptor. In some cases, TLR1 is also involved [88]. The interaction with the receptors triggers cascades of cellular signals within the cell, and the complex pathways culminate in a variety of host cell responses. Mycoplasmas and mycoplasmal LPP are known to activate the transcription factors NF-κB [74, 79] and AP-1 [1 4], via TLR-downstream cascades involving kinases (MAPKKKs-IKKs and MAPKKKs-MAPKKs-MAPKs). Known mycoplasma-affected target genes are mainly those responsible for proinflammatory proteins [4], and those involved in malignant cell transformation [72], with little information available on genes responsible for other proteins [53, 79, 80, 81].

5. Detecting mycoplasmas in cell cultures

The ubiquitous nature of mycoplasma in man, animals and the environment increases the likelihood of the introduction of these organisms into cell cultures or a manufacturing process. Currently, the recommended test requirements for biologics are as follows: (1) The master- and working cell seed banks must be free of mycoplasmas. (2) The product-harvest concentrates must be free of mycoplasmas. (3) All products produced in cell cultures, a generic term used for all tissue cells grown in vitro, must be tested. This includes viral vaccines (such as poliovirus, adenovirus, measles, rubella, mumps and rabies), monoclonal antibodies, immunological modifiers and cell-culture-derived blood products, such as tissue-type plasminogen and erythropoietin. Guidelines for mycoplasma testing of cell cultures and biologics is addressed in several international pharmacopoeias e.g., United States Pharmacopoeia, (USP 33/NF 28 <63>and <1226>, Mycoplasma tests, 2010); European Pharmacopoeia (EP 2.6.7., Mycoplasmas, 7th ed.; 2012); Japanese Pharmacopoeia (JP); Section 21 of the Code of Federal Regulations (CFR), International Conference on Harmonisation (ICH), and FDA- Points to Consider (PTC) documents. Several different approaches are being used to detect mycoplasmas in contaminated tissue cultures including the culture procedures, a variety of nonspecific procedures and the polymerase chain reactions (PCR).

5.1. Standard culture procedures

The culture procedures require that the tested material will be inoculated onto solid and liquid growth media capable of growing a variety of mycoplasma including aerobic, microaerophilic and anaerobic strains. Broth cultures are incubated and sub passaged to plate agar. After the required incubation period, the agar plates are observed microscopically for the presence of mycoplasma colonies [5]. The variation inherent in the complex media usually used for in vitro culture of mycoplasmas is due to batch variation in compounds such as sera, or yeast extract. Such variation makes the development of defined media attractive. However, a key problem has been the supply of lipids in an available, but non-toxic form, hence, defined artificial media have been developed for only a few species [1]. Most mycoplasmas produce microscopic (100 - 400 μm in diameter) colonies with a characteristic 'fried-egg' appearance, growing embedded in the agar, although some (e.g. *M. pulmonis*) may not grow completely embedded, and some freshly-isolated pathogens (e.g. *M. pneumoniae*) produce a more granular, diffuse colony-type. Since they usually grow embedded, mycoplasma colonies can be distinguished from other bacteria by: (1) specific colony shape; (2) being difficult to scrape from the agar surface. Mycoplasmas growing on agar can be identified more specifically by immunofluorescent procedures, using fluorophores conjugated to species-specific antibodies [4]. The traditional culture-based techniques are relatively sensitive, capable of detecting as few as 1-10 colony forming units of mycoplasmas and therefore are required by pharmacopoeias and regulatory authorities worldwide. Nonetheless, this procedure is time consuming requiring a minimum of 28 days to complete, costly and not sensitive to non-cultivable strains, therefore, the development of more accurate and faster techniques are needed to facilitate faster detection of a contaminating mycoplasma and more rapid corrective action.

5.2. Polymerase chain reaction (PCR)

PCR methodology has existed for decades, however conventional PCR and real-time PCR assays have only recently been considered for mycoplasma detection in cell cultures and biological products. These assays are often based on the amplification of conserved regions of the 16S rDNA [89, 90] or the spacer region between the 16S and 23S rDNA [91, 92]. The PCR approach is rapid (1-2 days), inexpensive, and independent of culture conditions. Specific oligonucleotide primers capable of amplifying the conserved regions and thus detecting DNA of multiple *Mollicutes* species while excluding other contaminating DNA are used in the PCR assays. In comparison to conventional PCR methods, real-time PCR assays are quicker, simpler, and more suitable for handling a large number of samples [93]. Nonetheless, some of the primers used are not entirely specific for *Mollicutes* [94, 95]. Thus, sequence homologies between *Mollicutes* spp. and *Chlamydia* spp. led to false-positive results in Chlamydial cell cultures tested for mycoplasma contamination with a commercial PCR kit [96].

Throughout the last decade, new PCR assays for mycoplasma detection, which appeared to resolve these issues, were described, while being sufficiently simple and inexpensive for routine use. For example, a PCR assay which applied readily available techniques in DNA extraction together with a modified single-step PCR using a primer pair that was homologous to a broad spectrum of mycoplasma species was proposed [97]. A high sensitivity and specificity for mycoplasma detection in cell production cultures was made possible through the combination of three key techniques: 8-methoxypsoralen and UV light treatment to decontaminate PCR reagents of DNA; hot-start Taq DNA polymerase to reduce nonspecific priming events; and touchdown PCR to increase sensitivity while also reducing nonspecific priming events. Another proposed PCR assay for mycoplasma detection was a sensitive two-stage PCR procedure which detected 13 common mycoplasmal contaminants [92]. For primary amplification, the DNA regions encompassing the 16S and 23S rRNA genes of 13 species were targeted using general mycoplasma primers. The primary PCR products were then subjected to secondary nested PCR, using two different primer pair sets, designed via the multiple sequence alignment of nucleotide sequences obtained from the 13 mycoplasmal species. The nested PCR, which generated DNA fragments of 165-353 bp, was found to be able to detect 1-2 copies of the target DNA, and evidenced no cross-reactivity with the genomic DNA of related microorganisms or of human cell lines, thereby confirming the sensitivity and specificity of the primers used.

Other studies showed that reverse transcription-PCR (RT-PCR) methods based on detection of the 16S rRNA, which is present in multiple (10^3–10^4) copies per bacterial cell [98, 99], are more sensitive than PCR detecting the 16S rDNA. Thus, a direct side-by-side comparison of RT-PCR and PCR targeting the 16S rRNA and the 16S rRNA gene, respectively, demonstrated that RT-PCR was able to provide up to a two-logarithm higher sensitivity of bacteria detection in comparison with the PCR-based assay [90, 100] and the sensitivity provided by RT-PCR is approaching the sensitivity of conventional microbiological culture methods [100]. Therefore, it was suggested that RT-PCR methods targeting the bacterial 16S or 23S rRNAs are having the real potential to provide the sensitivity of mycoplasma detection close to or even higher than that of conventional culture methods [101] .

Recently, the MycoTOOL PCR test kit from Roche (Roche, Diagnostic GmbH, Penzberg, Germany) was approved by the European Medicines Agency (EMEA) for release testing of pharmaceutical products. It is the first commercially available Mycoplasma PCR test that can replace traditional Mycoplasma tests (culture method as well as indicator cell culture method) during pharmaceutical production. In June 2009 the FDA approved the PCR concept of this test for seven commercial products from Genentech. Earlier, Bayer Health Care received approval for a pharmaceutical product from the EMEA and Japan's Ministry of Health, Labour and Welfare (MHLW) using the same PCR-based test concept. Guidelines describing acceptable protocols for specific PCR methods are provided by the EP and JP. The pharmacopoeias, PTC, and CFR protocols vary with their recommendations on how to conduct the PCR assays.

5.3. Indirect non-specific procedures

Some 'non-cultivable' mycoplasma strains cannot readily be grown on standard agar or broth-culture media [5], and cell-assisted culture is required for their isolation. Cell-culture systems are therefore a valuable ancillary tool for the isolation and detection of mycoplasmas and 'indicator cell culture' procedures using either VERO (African green monkey kidney), or NIH 3T3 cell cultures have been developed [102]. These cell lines are susceptible to infection by the majority of mycoplasmas and are therefore a reliable 'indicator' system for detecting mycoplasma infection. These approaches are particularly useful for the identification and detection of mycoplasmas that adhere to host-cell surfaces.

The indirect non-specific procedures require that the tested material will be inoculated directly onto tissue culture cover slips or flasks containing a monolayer of the indicator cells. The indicator cell culture inoculated with the tested material are than fixed and stained with DNA-binding fluorochromes using bisbenzimidazole (such as Hoechst or DAPI stains) [103].

Identification of contaminating mycoplasma is by visual observation via fluorescent microscopy. Mycoplasmas are detected by their characteristic particulate or filamentous pattern of bright fluorescence on the cell surface (Figure 4) and, if contamination is heavy, in surrounding areas. These procedures are suitable for use with either non-specific DNA stains for detecting mycoplasmas, or in conjunction with mycoplasma-speciation methods, such as by immunofluorescence procedures using species-specific polyclonal antisera, or monoclonal antibodies, conjugated with fluorescein or peroxidase [104]. A wide variety of luminol-dependent chemiluminescence and bioluminescent methods were described [5, 63].

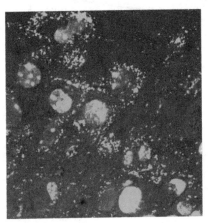

Figure 4. Mycoplasma contaminated eukaryotic cells stained with a fluorescent DNA stain.

Biochemical identification methods have also been in use [5, 78]. Procedures based on the comparative utilization of uridine versus uracil in contaminated versus mycoplasma-free cell cultures have been suggested [105]. Other methods are based on the detection of enzyme activity present in mycoplasmas, but absent, or minimal in uninfected cell cultures. The enzymic activities measured include: arginine deiminase [62]; thymidine, uridine, adenosine or pyrimidine nucleoside phosphorylase [102]; hypoxanthine or uracil phosphoribosyl transferase activities [106]. Positive results are based on arbitrary values, making low levels of mycoplasma contamination difficult to detect. Detection kit that provide a new, sensitive and rapid biochemical method was recently presented (Cambrex, Bio Science, Caravaggio, Bergamo, Italy). The test is based on a bioluminescent assay which can be assessed within 20 min for daily determination of the mycoplasma status of cell cultures. The performance sensitivity and specificity of the kit was evaluated and compared to the PCR/ELISA detection kit (Roche, Diagnostic GmbH, Penzberg, Germany) and the standard culture method [5]. Recently, a simple and inexpensive assay monitoring mycoplasma contamination, based on degradation of the Gaussia luciferase reporter in cell cultures was described [107]. This assay has been shown to be more sensitive for detecting mycoplasma contamination in seven different cell lines as compared to a commercially available bioluminescent assay [107]

6. Eliminating mycoplasmas from infected cultures

Ever since mycoplasma contamination of cell cultures was first reported, attempts have been made to develop methods for the elimination of mycoplasmas, including the use of antibiotics such as tetracycline, kanamycin, novobiocin, tylosin, gentamycin, doxycycline, thiayline and quinolones; surface-active agents; anti-mycoplasma antisera and prolonged heating treatments (40-42 °C) [63, 108]. Eliminating mycoplasmas by passage of a cell culture through nude mice [109] has been successful for some, but not all, mycoplasmas. An efficient procedure for eliminating mycoplasmas is based on the selective incorporation of 5-bromouracil (5-BrUra) into mycoplasmas, and the induction of breaks by light in the 5-BrUra-containing DNA [110]. The unusually high content of A+T makes the mycoplasma DNA an excellent candidate for the induction of breakage by the combined action of 5-BrUra, 33258-Hoechst and visible light [110]. Some of the elimination procedures may apply to some, but not all, mycoplasma species; some of them are laborious and/or time consuming. It was suggested, therefore, that whenever possible, the infected cell culture should be discarded and replaced with a mycoplasma-free culture [108]. When the cell culture is irreplaceable, the use of antibiotic mixtures, are the commonest approaches. One has to keep in mind that cell-culture contaminants that have been continuously exposed to antibiotics develop resistance to the drug, and antibiotic-resistant strains have been isolated for most *Mycoplasma* species tested. Treatment may also induce the selection of a subpopulation of cells and the treated cell culture may differ in its characteristics from the original culture.

Among the antibiotics that were shown to have strong anti-mycoplasma properties are different inhibitors of protein synthesis mainly tetracyclines or macrolides as well as quinolones [111]. The target enzymes of quinolones are considered to be DNA gyrase and topoisomerase IV which are essential enzymes for controlling the topological state of DNA in DNA replication and transcription . Most recently the quinolone garenoxacin was found to be a most valuable quinolone in the elimination *M. pneumoniae* [112].

The addition of antibiotics to the culture medium during a limited period of time (1-3 wk) is a simple, inexpensive, and very practical approach for decontaminating continuous cell lines. BM-cyclin (trade name of Roche, Mannheim, Germany), a combination of tiamulin and minocycline (both inhibiting protein synthesis), was introduced by Jung et al. [113] who show that three cycles of treatment of a contaminated cell culture with BM-cyclin I (containing the macrolide tiamulin) at a final concentration of 10 µg/ml for 3 days followed by BM-cyclin II (containing the tetracycline minocycline) at a concentration of 5 µg/ml for 4 days completely eradicated mycoplasmal infection from cultured cells [113].

Uphoff and Drexler [111, 114] examined the effectiveness of several quinolones and BM-cycline protocols. The contaminated cell cultures were exposed to one of the following five antibiotic regimens: mycoplasma removal agent (MRA, quinolone; a 1-wk treatment), enrofloxacin (quinolone; 1 wk), sparfloxacin (quinolone; 1 wk), ciprofloxacin (quinolone; 2 wk), and BM-Cyclin (alternating tiamulin and minocycline; 3 wk). The mycoplasma infection was permanently eliminated by the various antibiotics in 66-85% of the cultures treated. Mycoplasma resistance was seen in 7-21%, and loss of the culture as a result of cytotoxically caused cell death occurred in 3-11% of the cultures treated [111, 114].

Recently, MycoZap (trade name of Lonza, Verviers, Belgium) treatment has been introduced as a new therapeutic tool able to overcome the eukaryotic cytotoxicity of fluoroquinolones and BM-Cyclins [115]. MycoZap kit (Lonza, Verviers, Belgium) includes a combination of patented antibiotic and antimetabolic agents. An evaluation of the MycoZap kit performance was recently presented by Mariotti et al., [116] who exposed mycoplasma contaminated cells to the MycoZap protocol and compared the results obtained to the eradication efficiency of enrofloxacin (Fluka, Bio-Chemika, Missouri, USA), MRA (Euroclone, Lugano, Switzerland), ciprofloxacin and the BM-Cyclin protocol. Treatment of contaminated cell cultures by MycoZap, MRA, ciprofloxacin, enrofloxacin and BM-cycline, eliminated mycoplasma infection by 46%, 29%, 43%, 40% and 57% respectively. The use of an eradication mixture based on a combination of the antibiotics BM-Cyclins, ciprofloxacin, enrofloxacin and MRA was able to clean 88.6% of the infected cultures, whereas the addition of MycoZap to the eradication mixture resulted in the eradication of mycoplasmas from 100% of the contaminated cell cultures [116].

7. Conclusions

Mycoplasmas are shown to cause various alterations in cultured cells. As described above, some alterations are due to direct effects on the cells by mycoplasma components, and other alterations are due to indirect effects, via inducing the host cell to alter its gene and protein expression and activity. It is important to emphasize the fact that mycoplasmal-altered cell phenotype and function is often observed in specific types of cells under special conditions, e.g., when the cultured cells are exposed to certain agents. The detection of mycoplasma contamination, and the identification of the factors and pathways involved in the mycoplasmal effects are thus of utmost importance in handling cultured cells, including using stem cells for differentiation to specific tissues.

Author details

Shlomo Rottem and Jonathan D. Kornspan
Department of Microbiology and Molecular Genetics, IMRIC,
The Hebrew University-Hadassah Medical School, Jerusalem, Israel

Nechama S. Kosower
Department of Human Molecular Genetics and Biochemistry, Sackler School of Medicine, Tel-Aviv
University, Ramat-Aviv, Tel-Aviv, Israel

Acknowledgement

Figure 1A is a courtesy of S. Razin, The Hebrew University-Hadassah Medical School and Figure 1B is a courtesy of A. M. Collier, The University of North Carolina School of Medicine. We would like to thank M. Tarshis for the confocal microscopy.

8. References

[1] Razin S, Yogev D, Naot Y (1998) Molecular biology and pathogenicity of mycoplasmas. Microbiol. Rev. 63: 1094-1156.

[2] Fraser CM, et. al. (1995) The minimal gene complement of *Mycoplasma genitalium*. Science 270: 397-403.

[3] Maniloff J (1996) The minimal gene genome: "on being the right size." Proc. Natl. Acad. Sci. USA 93: 10004-10006.

[4] Rottem S (2003) Interaction of mycoplasmas with host cells. Physiol. Rev. 83: 417-432.

[5] Barile MF, Rottem S (1993) Mycoplasmas in cell cultures. In: Kahane I, Adoni A, editors. Rapid diagnosis of mycoplasmas. New York: Plenum Press. pp. 155-193.

[6] Kornspan JD, Tarshis M, Rottem S (2010) Invasion of melanoma cells by *Mycoplasma hyorhinis*: enhancement by protease treatment. Infect. Immun. 78: 611-617.

[7] Shaw JH, Falkow S (1988) Model for invasion of human tissue culture cells by *Neisseria gonorrhoeae*. Infect. Immun. 56: 1625-1632.

[8] Razin S, Jacobs E (1992) Mycoplasma adhesion. J. Gen. Microbiol. 138: 407-422.

[9] Roberts DD, et.al. (1989) Sialic acid-dependent adhesion of *Mycoplasma pneumoniae* to purified glycoproteins. J. Biol. Chem. 264: 9289-9293.

[10] Geary SJ, Gabridge MG (1987) Characterization of a human lung fibroblast receptor site for *Mycoplasma pneumoniae*. Isr. J. Med. Sci. 23: 462-468.

[11] Geary SJ, et.al. (1990) Identification of mycoplasma binding proteins utilizing a 100 kilodalton lung fibroblast receptor. J. Rec. Res. 9: 465-478.

[12] Krivan Hc, et. al. (1989) Adhesion of *Mycoplasma pneumoniae* to sulfated glycolipids and inhibition by dextran sulfate. J. Biol. Chem. 264: 9283-9288.

[13] Debey MC, Ross RF (1994) Ciliostasis and loss of cilia induced by *Mycoplasma hyopneumoniae* in porcine tracheal organ cultures. Infect. Immun. 62: 5312-5318.

[14] Almagor M, et. al. (1986) Protective effects of the glutathione redox cycle and vitamin E on cultured fibroblasts infected by *Mycoplasma pneumoniae*. Infect. Immun. 52: 240-244.

[15] Shibata KI, Sasaki T, Watanabe T (1995) AIDS-Associated mycoplasmas possess phospholipases C in the membrane. Infect. Immun. 63: 4174-4177.

[16] Rosenshine I, Finlay BB (1993) Exploitation of host signal transduction pathways and cytoskeletal functions by invasive bacteria. BioEssays 15: 17-24.

[17] Salman M, Rottem S (1995) The cell membrane of *Mycoplasma penetrans*: Lipid composition and phospholipase A1 activity. Biochim. Biophys. Acta. 1235: 369-377.

[18] Salman M, Borkovsky Z, Rottem S (1998) *Mycoplasma penetrans* invasion of Molt-3 lymphocytes induces changes in the lipid composition of host cells. Microbiology 144: 3447-3454.

[19] Lo SC (1992) Mycoplasmas in AIDS. In: Maniloff J, McElhaney RN, Finch LR, Baseman JB editors. Mycoplasmas: Molecular Biology and Pathogenesis. Washington, D.C.: American Society Microbiology. pp. 525-548.

[20] Lo SC, et. al. (1993) Adhesion onto and invasion into mammalian cells by *Mycoplasma penetrans*: a newly isolated mycoplasma from patients with AIDS. Mod. Pathol. 6: 276-280.

[21] Andreev J, et. al. (1995) Invasion of HeLa cells by *Mycoplasma penetrans* and the induction of tyrosine phosphorylation of a 145 kDa host cell protein. FEMS. Microbiol. Letts. 132: 189-194.

[22] Borovsky Z, et. al. (1998). *Mycoplasma penetrans* invasion of HeLa cells induces protein kinase C activation and vacuolation in the host cells. J. Med. Microbiol. 47: 915-922.

[23] Giron JA, Lange M, Baseman JB (1996) Adherence, fibronectin binding, and induction of cytoskeleton reorganization in cultured human cells by *Mycoplasma penetrans*. Infect. Immun. 64: 197-208.

[24] Chausee MS, Cole R, Van Putten JPM (2000) Streptococcal erythrogenic toxin B abrogates fibronectin dependent internalization of *Streptococcus pyogenes* by cutured mammalian cells. Infect. Immun. 68: 3226-3232.

[25] Yavlovich A, Katzenell A, Rottem S (2007) Binding of host extracellular matrix proteins to *Mycoplasma fermentans* and its effect on adherence to, and invasion of HeLa cells. FEMS Microbiol. Letts. 266: 158-162.

[26] Yavlovich A, Higazi AR, Rottem S (2001) Plasminogen binding and activation by *Mycoplasma fermentans*. Infect. Immun. 69: 1977-1982.

[27] Yavlovich, A., et. al. (2004) *Mycoplasma fermentans* binds to and invades HeLa cells: Involvement of plasminogen and urokinase. Infect. and Immun. 72, 5004-5011.

[28] Rao JS (2003) Molecular mechanisms of glioma invasiveness: the role of proteases. Nat. Rev. Cancer 3:489-501.

[29] Balish MF, Krause DC (2002) Cytadherence and the cytoskeleton. In: Razin S, Herrmann R, editors. Molecular Biology and Pathogenicity of Mycoplasmas. New York: Kluwer Academic/Plenum. pp. 491-518.

[30] Jensen JG, Blom J, Lind K (1993) Intracellular location of *Mycoplasma genitalium* in cultured Vero cells as demonstrated by electron microscopy. Int. J. Path. 75: 91-98.

[31] Winner F, Rosengarten R, Citti C (2000) In vitro cell invasion of *Mycoplasma gallisepticum*. Infect. Immun. 68: 4238-4244.

[32] Zucker-Franklin D, Davidson M, Thomas L (1966) The interaction of mycoplasmas with mammalian cells, HeLa cells, neutrophiles, and eosinophils. J. Exp. Med. 124: 521-532.

[33] Taylor-Robinson D, et. al. (1991) Intracellular location of mycoplasmas in cultured cells demonstrated by immunocytochemistry and electron microscopy. Int. J. Exp. Pathol. 72: 705-714.

[34] Elsinghorst EA (1994) Measurement of invasion by gentamicin resistance. Methods Enzymol. 236: 405-420.

[35] Finlay BB, Ruschkowski S, Dedhar S (1991) Cytoskeletal rearrangements accompanying Salmonella entry into epithelial cells. J. Cell Sci. 99: 283-296.

[36] Baseman JB, Tully JG (1997) Mycoplasmas: sophisticated, reemerging, and burdened by their notoriety. Emerg. Infect. Dis. 3: 21-32.

[37] Feng SH, et. al. (1999) Mycoplasmal infections prevent apoptosis and induce malignant transformation of interleukin-3-dependent 32D hematopoietic cells. Mol. Cellular Biol. 19: 7995-8002.

[38] Stadtlander CT, et. al (1993) Cytopathogenicity of *Mycoplasma fermentans* (including strain incognitos). Clin. Infect. Dis. 17: 289-301.

[39] Oelschlaeger TA, Kopecko DJ (2000) Microtubule dependent invasion pathways to bacteria. In: Oelschlaeger TA, Hacker J, editors. Bacterial Invasion into Eukaryotic Cells.

Subcellular Biochemistry, Volume 33. New York: Kluwer Academic/Plenum Publishers. pp. 3-19.

[40] Salman M, et. al. (1994) Membrane lipids of Mycoplasma fermentans. FEMS Microbiol. Lett. 123: 255-260.

[41] Rottem S, Naot Y (1998) Subversion and exploitation of host cells by mycoplasmas. Trends. Microbiol. 6: 436-440.

[42] Tarshis M, Salman M, Rottem S (1993) Cholesterol is required for the fusion of single unilamellar vesicles with *M. capricolum.* Biophys. J. 64: 709-715.

[43] Cullis PR, Hope MJ (1988) Lipid polymophism, lipid asymmetry and membrane fusion. In: Ohki S, Doyle D, Flanagan TD, Hui SW, Mayhew E, editors. Molecular Mechanisms of Membrane Fusion. New York: Plenum Press. pp. 37-51.

[44] Lucy J (1970) The fusion of biological membranes. Nature 227: 814-817.

[45] Siegel DP (1999) Energetics of intermediates in membrane fusion: comparison of stalk and inverted micellar intermediate structures. Biophys. J. 76: 291-313.

[46] Ben-menachem G, Zähringer U, Rottem S (2001) The phosphocholine motif in membranes of *Mycoplasma fermentans* strains. FEMS Microbiol. Letts 199: 137-141.

[47] Deutsch J, Salman M, Rottem S (1995) An unusual polar lipid from the cell membrane of *Mycoplasma fermentans.* Eur J Biochem 227: 897-902.

[48] Zahringer U, et. al. (1997) Primary structure of a new phosphocholine-containing glycoglycerolipid of *Mycoplasma fermentans.* J. Biol. Chem. 272: 26262-26270.

[49] Wagner F, et. al. (2000) Ether lipids in the cell membrane of *Mycoplasma fermentans.* Eur. J. Biochem. 267: 6276-6286.

[50] Franzoso G, et. al. (1992) Fusion of *Mycoplasma fermentans,* strain incognitus, with T-lymphocytes. FEBS Lett. 303: 251-254.

[51] Dimitrov DS, et. al. (1993) *Mycoplasma fermentans,* incognitus strain, cells are able to fuse with T-lymphocytes. Clin. Infect. Dis. 17: S305-S308.

[52] Shibata KI, et. al. (1994) Acid phosphatase purified from *Mycoplasma fermentans* has protein tyrosine phosphatase-like activity. Infect. Immun. 62: 313-315.

[53] Sokolova IA, Vaughan AT, Khodarev NN (1998) Mycoplasma infection can sensitize host cells to apoptosis through contribution of apoptotic-like endonuclease(s). Immunol. Cell. Biol. 76: 526-534.

[54] van Rijn J, et. al. (2004) Induction of hyperammonia in irradiated hepatoma cells: recapitulation and possible explanation of the phenomenon. Brit. J. Cancer 91: 150-152.

[55] van den Voorde J, et. al. (2012) Characterization of pyrimidine nucleoside phosphorylase of *Mycoplasma hyorhinis*: implications for the clinical efficacy of nucleoside analogues. Biochem. J. (In press).

[56] Gerlic M, et. al. (2007) The inhibitory effect of *Mycoplasma fermentans* on tumor necrosis factor (TNF)-alpha-induced apoptosis resides in the membrane lipoproteins. Cell Microbiol. 9: 142-153.

[57] Obara H, Harasawa R (2010) Nitric oxide causes anoikis through attenuation of E-cadherin and activation of caspase-3 in human gastric carcinoma AZ-521 cells infected with *Mycoplasma hyorhinis.* J. Vet. Med. Sci. 72: 869-874.

[58] Dusanic D, et. al. (2012) *Mycoplasma synoviae* induces upregulation of apoptotic genes, secretion of nitric oxide and appearance of an apoptotic phenotype in infected chondrocytes. Veterinary Res. 43: 7-20.

[59] Zhang S, Lo SC (2007) Effect of mycoplasmas on apoptosis of 32D cells is species-dependent. Curr. Microbiol. 54: 388-395.

[60] Himmelreich R, et. al (1997) Comparative analysis of the genomes of the bacteria *Mycoplasma pneumoniae* and *Mycoplasma genitalium*. Nucleic Acids Res. 25: 701-712.

[61] Rechnitzer H., et. al. (2011) Genomic features and insights into the biology of *Mycoplasma fermentans* Microbiology 157: 760–773.

[62] Schimke RT, Barile MF (1963) Arginine breakdown in mammalian cell culture contaminated with pleuropneumonia-like organisms (PPLO). Exp. Cell Res. 30: 593-596.

[63] Rottem S, Barile MF (1993) Beware of mycoplasmas. Trends. Biotechnol. 11: 143-151.

[64] Ben-menachem G, et. al. (2001) Choline-difficiency induced by *Mycoplasma fermentans* enhances apoptosis of rat astroctes. FEMS Microbiol. Letts. 201: 157-162.

[65] Zeisel SH, Blusztajn JK (1994) Choline and human nutrition. Annu. Rev. Nutr. 14: 269-296.

[66] Bendjennat M, et. al. (1999) Role of *Mycoplasma penetrans* endonuclase P40 as a potential pathogenic determinant. Infect. Immun. 67: 4456-4462.

[67] Minion FC, et. al. (1993) Membrane-associated nuclease activities in mycoplasmas. J. Bacteriol. 175: 7842-7847.

[68] Paddenberg R, et. al. (1996) Internucleosomal DNA fragmentation in cultured cells under conditions reported to induce apoptosis may be caused by mycoplasma endonucleases. Eur. J. Cell Biol 71: 105-119.

[69] Sokolovai A, Vaughan AT, Khodarev NN (1998) Mycoplasma infection can sensitize host cells to apoptosis through contribution of apoptotic-like endonuclease(s). Immunol. Cell Biol. 76: 526-534.

[70] Tsai S, et. al. (1995) Mycoplasma and oncogenesis persistent infection and multistage malignant transformation. Proc. Natl. Acad. Sci. USA 92: 10197-10201.

[71] Zhang B, et. al. (1997) High-level expression of H-ras and c-myc onconogenes in mycoplasma-mediated malignant cell tranformation. Proc. Soc. Exp. Biol. Med. 214: 359-366.

[72] Zhang S, Tsai S, Lo SC (2006) Alteration of gene expression profiles during mycoplasma-induced malignant cell transformation. BMC Cancer 6: 116-126.

[73] Jiang S, et. al. (2008) Mycoplasma infection transforms normal lung cells and induces Bone Morphogenetic Protein 2 expression by post-transcriptional mechanisms. J. cellul. Biochem. 104: 580-594.

[74] Logunov DY, et. al. (2008) Mycoplasma infection suppresses p53, activates NF-κB and cooperates with oncogenic Ras in rodent fibroblast transformation. Oncogene 27: 4521-4531.

[75] Gong M, et. al. (2008) p37 from *Mycoplasma hyorhinis* promotes cancer cell invasiveness and metastasis through activation of MMP-2 and followed by phosphorylation of EGFR. Mol. Cancer Ther. 7: 530-537.

[76] Cole BC, et. al. (2005) Isolation and partial purification of macrophage-and dentritic cell-activating components from *Mycoplasma arthritidis:* association with organism virulence and involvement with Toll-like receptor 2. Infect. Immun. 73: 6039-6047.

[77] Reyes L, et. al. (1999) Effects of *Mycoplasma fermentans* incognitus on differentiation of THP-1 cells. Infect. Immun. 67: 3188-3192.

[78] Quah BJC, O'Neill HC (2007) Mycoplasma contaminants present in exosome preparations induce polyclonal B cell response. J. Leukoc. Biol. 82: 1070-1082.

[79] Elkind E, et. al. (2012) Calpastatin upregulation in *Mycoplasma hyorhinis*-infected cells is promoted by the mycoplasma lipoproteins via the NF-κB pathway. Cell. Microbiol. 14: 840-851.

[80] Dong L, et. al. (1999) Transcriptional activation of mRNA of intercellular Adhesion Molecule 1 and induction of its cell surface expression in normal human gingival fibroblasts by *Mycoplasma salivarium* and *Mycoplasma fermentans*. Infect. Immun. 67: 3061-3065.

[81] Kraft M, et al. (2008) *Mycoplasma pneumoniae* induces airway epithelial cell expression of MUC5AC in asthma. Eur. Respir. J. 31: 43-46.

[82] Goll DE, et. al. (2003) The calpain system. Physiol. Rev. 83: 731-801.

[83] Manischewitz JE, Young BG, Barile MF (1975) The effect of mycoplasmas on replication and plaquing ability of Herpes simplex virus. Proc. Soc. Exp. Biol. Med. 148: 859-863.

[84] Singer SH, et. al. (1970) Effect of mycoplasmas on vaccinia virus growth: requirement for arginine. Proc. Soc. Exp. Biol. Med. 133: 1439-1442.

[85] Harper DR, et. al. (1988) Reduction in immunoreactivity of varicella-zoster virus proteins induced by mycoplasma contamination. J. Virol. Methods 20: 65-72.

[86] Beck J, et. al. (1982) Induction of interferon by mycoplasmas in mouse spleen cell cultures. J. Interferon Res. 2: 31-36.

[87] McGarrity GJ, Kotani H (1985) Cell Culture Mycoplasmas. In: Razin S, Barile MF, editors. The Mycoplasmas, Vol. IV. New York & London: Academic Press. pp. 353-390.

[88] Zuo LL, Wu YM, You XX (2009) Mycoplasma lipoproteins and Toll-like receptors J. Zhejiang Univ. Sci. B 10: 67-76.

[89] Jurstrand M, et. al. (2005) Detection of *Mycoplasma genitalium* in urogenital specimens by real-time PCR and conventional PCR assay. J. Med. Microbiol. 54: 23–29.

[90] van Kuppeveld FJ, et. al. (1993) Genus- and species-specific identification of mycoplasmas by 16S rRNA amplification. Appl. Environ. Microbiol. 59: 655.

[91] Harasawa R, et. al. (2005) Rapid detection and differentiation of the major mycoplasma contaminants in cell cultures using real-time PCR with SYBR Green I and melting curve analysis. Microbiol. Immunol. 49: 859–863.

[92] Sung H, et. al. (2006) PCR-based detection of mycoplasma species. J. Microbiol. 44: 42–49.

[93] Störmer M, et. al. (2009) Broad-range real-time PCR assay for the rapid identification of cell-linecontaminants and clinically important mollicute species. Int. J. Med. Microbiol. 299: 291-300.

[94] Waites KB, Brenda K, Schelonka RL (2005) Mycoplasmas and ureaplasmas as neonatal pathogens. Clin. Microbiol. Rev. 18: 757–789.

[95] Wang H, et. al. (2004) Simultaneous detection and identification of common cell culture contaminant and pathogenic *Mollicutes* strains by reverse line blot hybridization. Appl. Environ. Microbiol. 70: 1483–1486.

[96] Maass V, et. al. (2011) Sequence homologies between *Mycoplasma* and *Chlamydia* spp. lead to false-positive results in chlamydial cell cultures tested for mycoplasma contamination with a commercial PCR assay. J. Clin. Microbiol. 49: 3681-3682.

[97] Eldering JA, et. al. (2004) Development of a PCR method for mycoplasma testing of Chinese hamster ovary cell cultures used in the manufacture of recombinant therapeutic proteins. Biologicals 32: 183-193.

[98] Waters AP, McCuthan TF (1990) Ribosomal RNA: nature's own polymerase-amplified target for diagnosis. Parasitol. Today 6: 56–59.

[99] Ortiz JO, et. al. (2006) Mapping 70S ribosomes in intact cells by cryoelectron tomography and pattern recognition. J. Struct. Biol. 156: 334–341.

[100] Matsuda K, et. al. (2007) Sensitive quantitative detection of commensal bacteria by rRNA-targeted reverse transcription-PCR. Appl. Environ. Microbiol. 7: 32–39.

[101] Peredeltchouk M, et. al. (2011) Detection of mycoplasma contamination in cell substrates using reversetranscription-PCR assays. J. Appl. Microbiol. 110: 54-60.

[102] Spierenburg GT, et. al. (1988) Indicator cell lines for the detection of hidden mycoplasma contamination, using an adenosine phosphorylase screening test. J. Immunol. Methods 114: 115-119.

[103] Chen TR (1977) In situ detection of mycoplasma contamination in cell cultures by fluorescent Hoechst 33258 stain. Exp. Cell Res. 104: 255-262

[104] Freiberg EF, Masover GK (1990) Mycoplasma detection in cell culture by concomitant use of bisbenzamide and fluoresceinated antibody. In vitro Cell Dev. Biol. 26: 585-588.

[105] Perez AG, et. al. (1972) Altered incorporation of nucleic acid precursors by mycoplasma-infected mammalian cells in culture. Exp. Cell Res. 70: 301-310.

[106] Mariotti E, et. al. (2008) Rapid detection of mycoplasma in continuous cell lines using a selective biochemical test. Leuk. Res. 32: 323-326.

[107] Degeling MH, et. al. (2012) Sensitive assay for mycoplasma detection in mammalian cell culture Anal. Chem. 84: 4227–4232.

[108] McGarrity GY, Kotani H, Bulter GH (1992) Mycoplasmas and tissue culture cells, In: Maniloff J, McElhaney RN, Finch LR, Baseman JB, editors. Mycoplasmas: Molecular Biology and Pathogenesis. Washington, D.C.: American Society Microbiology. pp. 445-454.

[109] van Diggelen OP, Shin SI, Phillips DM (1977) Reduction in cellular tumorigenicity after mycoplasma infection and elimination of mycoplasma from infected cultures by passage in nude mice. Cancer Res. 37: 2680-2687.

[110] Marcus M, et. al. (1980) Selective killing of mycoplasmas from contaminated mammalian cells in cell cultures. Nature 285: 659-661.

[111] Uphoff CC, Drexler HG (2005) Eradication of mycoplasma contaminations. Methods Mol. Biol. 290: 25-34.

[112] Nakatani M, et. al. (2012) Inhibitory activity of garenoxacin against DNA gyrase of *Mycoplasma pneumoniae*. J. Antimicrob. Chemother. (in press)

[113] Jung H, et. al. (2003) Detection and treatment of mycoplasma contamination in cultured cells. Chang Gung Med. J. 26: 250-258.

[114] Uphoff CC, Drexler HG (2002) Comparative antibiotic eradication of mycoplasma infections from continuous cell lines. In Vitro Cell Dev. Biol. Anim. 38: 86-89.

[115] Uphoff CC, Drexler HG (2011) Elimination of mycoplasmas from infected cell lines using antibiotics. Methods Mol. Biol. 731: 105-114.

[116] Mariotti E, et. al. (2012) Mollicutes contamination: a new strategy for an effective rescue of cancer cell lines. Biologicals 40:88-91.

Chapter 3

Cell Handling and Culture Under Controlled Oxygen Concentration

Satoru Kaneko and Kiyoshi Takamatsu

Additional information is available at the end of the chapter

1. Introduction

The term "cell culture" refers to the in vitro proliferation of cells and tissues and involves the use of either primary cells typically having a finite life span in culture or continuous cell lines that are abnormal and immortalized. In the last several decades, efforts have been mainly focused on the sub-culturing of established cell lines and on optimizing culture conditions, including selection of appropriate culture medium, in order to achieve rapid cell growth. The conventional CO_2 incubator only provides minimum requirements for keeping cells alive in a culture environment; pH and temperature are held at 7.4 and 37°C, respectively, and sub-saturated humidity is maintained to avoid evaporative condensation of the culture medium. To avoid contamination of the culture medium from aerosol bacteria, the conventional clean bench removes particles from the working atmosphere by continuous displacement of ambient air that passes through a high-efficiency particle (HEPA) filter, whereas the medium exposed to ambient air quickly discharges dissolved CO_2. In vivo, the pH of blood is strictly maintained at ~ pH 7.4 by means of physiological buffer systems, whereas the pH of culture medium mainly depends on a sodium bicarbonate/CO_2 buffer, which is adjusted to a pH of 7.4 in an atmosphere of 5% CO_2; the pKa is 6.1, and the buffering capacity at pH 7.4 is very weak. The pH and temperature are generally uniform in almost all organs, tissues, and cells in living mammals; however, the dissolved oxygen or oxygen partial pressure (PO_2) in various organs and tissues is generally much lower than that in the ambient air (159 mmHg); it decreases to 100 mmHg and 25 mmHg or less in arterial blood and in the periphery, respectively. Although oxygen is essential to produce ATP through the tricarboxylic acid cycle, it is quite toxic at high concentrations [1-3]. Tissues and cells in body fluids are protected from reactive oxygen species (ROS) by multiple physiological anti-oxidant systems, whereas those in artificial culture medium lack an extracellular protective system and are exposed to high levels of O_2. Numerous studies have cited a variety of harmful effects of ROS, such as lipid and protein peroxidation as well as membrane and DNA damage [4-8]. Although lower PO_2 implies lower production of ROS, it also implies hypoxia that can damage various cellular functions [9, 10]. The recent advances in tissue engineering focus on clinical applications of cultured cells in regenerative medicine. When primary cells and stem cells are retrieved from human tissue, their original physicochemical environments and metabolic features are quite different from each other. Moreover, induction of differentiation stimulates changes in gene expression over a time course, which may affect cellular metabolism. The above-mentioned point of view suggests that ranges in cell viability, in terms of PO_2 and tolerance for ROS (which need to be controlled for normal proliferation and prevention of malignant transformation), may be quite narrow for primary cells in comparison to cell lines. The stringent control of oxygen during bioprocessing is undoubtedly important; however, the exact influence of PO_2 and ROS is not completely understood because past experimental data have been obtained by using conventional culture apparatuses or their improved models. Although the performance of apparatuses has already been proven by sub-culturing of established cell lines, they have not been designed for stringent control of oxygen throughout the culture period. Every time the incubator door is opened, the O_2 environment is quickly lost and requires a long time to recover. To clarify the net influence of PO_2 and ROS, it is essential to develop advanced equipment that can provide a stringent control of oxygen around the pericellular environment throughout the culture period.

2. Delivery of human babies by assisted reproductive technology: Elementary regenerative medicine involving the transplantation of cultured embryos

In the past 30 years, researchers have made significant progress in the field of clinical reproductive medicine through assisted reproductive technology (ART). In 1978, the first human baby was born through in vitro fertilization-embryo transfer. To date, more than a million children have already been born with the help of ART. Fertilization through intra-cytoplasmic sperm injection (ICSI) [11] is becoming increasingly popular, and it now accounts for more than half of clinical ART cases worldwide. ART is the first example of large-scale clinical application of regenerative medicine to cultured human stem cells, which involves in vitro fertilization of gametes (primary stem cells) and subsequent culture of early embryos up to the blastocyst stage, followed by transfer into the uterus. Some recent cohort studies could not deny the possibility of birth defects in babies who were delivered as a consequence of ART [12-14], although ART is recognized as an elementary clinical regenerative medicine that makes use of native stem cells. We point out two major issues in this regard: one is quality control of the gametes, and the other is quality assurance of the culture environments. It is well known that human ejaculate contains a heterogeneous sperm population that possesses a variety of abnormalities. ICSI is a technique mainly used in male infertility, which occurs as a result of dysfunction of spermatogenesis and is accompanied by various functional deteriorations in the sperm. Nuclear damage to human sperm, in particular, DNA fragmentation as a consequence of double-strand breaks, has attracted attention. If a sperm with damaged DNA is incorporated into the embryonic genome, it may lead to sperm-derived chromosomal aberrations [15], which may in turn result in higher miscarriage rates [16] and an increased risk of pregnancy loss [17]. The resultant aberrations can also be potentially inherited through the germ line by future generations [18-20]. Several studies have reported that the rate of DNA-damaged sperm increases in infertile men with poor semen quality, who are the primary candidates for ICSI [21]. Although the techniques in clinical ICSI are well established, the sperm is selected merely based on motility and gross morphology, as observed under a microscope, and there are no validated methods to address and assure sperm nuclear DNA integrity.

There is a concern about higher malformations resulting from ICSI cycles, due to the possibility of iatrogenic transmission of genetic abnormalities to the offspring [14, 22, 23]. Studies comparing ART cycles and natural births suggest that infants conceived by IVF / ICSI techniques have three times a risk of a congenital heart defect [24] as well as a higher risk of autosomal and gonosomal aneuploidies [25]. It still remains unclear whether culture environments provided by conventional culture apparatus or their improved models have been responsible for the results of various cohort analyses. In general, we have to consider the heterogeneity of the cell population at the start of culture as well as some transformation during the culture. The lumen of the fallopian tube, where the oocyte fertilizes with the sperm, shows very low PO_2 [26], contrary to the endometrium at the implantation phase, which shows thickening with increased blood flow; thus, the implanting blastocyst is exposed to higher PO_2. During this one-week trip in the oviduct, the embryo undergoes early development in a PO_2 gradient. To determine optimal physicochemical environment for primary and stem cells, including the embryo, one has to pay attention to complicated cross-interactions between the atmosphere, especially PO_2, and the composition of culture medium: ATP production is influenced by peri- and intra-cellular PO_2 as well as energy sources in the medium. Even if PO_2 is kept low during cell culture, the handling of cells in a clean bench is critical, while temperature, PO_2, and pH of the medium are dramatically changed. Tolerances to such parameters varies quite differently among cell types. For example, some neuronal cells have a low threshold for oxygen toxicity, and exposure of these cells to ambient air in a clean bench induces apoptosis [3]. Such cells have to be treated in an enclosed space filled with low-oxygen gas mixtures. In contrast, some cells can readily induce apoptosis under low PO_2 conditions [9]. It is well-known that long-term subculture of cell lines induces some genetic transformations. Some researchers [27-31] have proposed that this phenotypic variability might originate from epigenetic alterations, and the methylation profiles of stem cell lines are fundamentally changed during subculture, thus complicating their use in basic and clinical research. Several reports have also discussed the epigenetics of early development [32] and the genetic and epigenetic features of children delivered through ICSI [33]. Kohoda et al. suggested that ICSI induces transcriptome perturbation [34]. To ensure the reliability of clinical embryo cultures, or in general terms, clinical cell cultures, as a premise for human implantation, we have to recognize the complicated cross-interactions of gas phase with composition of the culture medium, cell features, and their heterogeneity with regard to genetic and epigenetic regulation. Numerous reports have emphasized that reducing PO_2 during in vitro cultures increases the proportion of blastocyst formation in mice [35-37], hamsters [38], sheep [39], and cattle [40].

Other studies found no clear effect in mice [41]. As mentioned above, PO_2 and ROS might be essential parameters at least in early embryogenesis [42]. The discrepancies in the results of the above-mentioned studies may be partially explained by differences in culture hardware as well as culture methods: for example, oxygen tension in droplets of medium under oil will be less than those without an oil overlay [43]. Adding EDTA to culture medium increased the proportions of mouse [41, 44] and cattle [40] embryos that developed to blastocysts. Chelating transition metals such as zinc, iron, and copper may prevent chemical reactions that generate harmful oxygen radicals [45]. Because oviductal oxygen tension is less than atmospheric levels [26], mammalian embryos may be protected from oxidative stress in vivo in part by a relatively low oxygen tension in the oviduct [46]. The influence of low PO_2 and hypoxic culture conditions on some cellular functions has also been studied in somatic cells. When BeWo cells, an in vitro model of human trophoblasts, were cultured in 2% O_2, reverse-transcriptase polymerase chain reaction (RT-PCR) indicated increased transcription of the organic cation transporter (OCTN2) gene compared to that observed under 20% O_2 [47]. Hirao et al. ([48] observed that MC3T3-E1 cells and calvariae from 4-day-old mice cultured in 5% or 20% O_2 conditions showed osteoblastic differentiation and subsequent transformation to osteocytes, which was promoted by low PO_2. The importance of a lower PO_2 environment was a cited factor; however, excessively low PO_2 are also important to consider for cellular growth, differentiation, gene expression, phenotype manipulation, epigenetics, and moreover, for survival. We consider it essential to determine the narrow range between hyperoxia and hypoxia, but not to overestimate the benefit of lower PO_2.

3. Individual cell culture systems in a disposable capsule with controlled atmosphere

As will be described later (Figs. 7 and 8), established cell lines that adapt to 5.0% CO_2-air often tolerate prolonged changes in pH and PO_2. In contrast, clinical cultures of primary and stem cells used for human transplantation demands rigorous duplication of in vivo environments because it is of prime importance for maintaining normality or to minimize phenotypic changes within the cells. We have previously established an individual cell culture system that emphasizes the precise control of oxygen concentration and quick recovery from disturbances (Figs. 1 and 2) [49]. As shown in Fig. 1, the culture bath has an aluminum block with 16 wells for heat storage, and the block and inner space are kept at 37.0 °C by a temperature sensor. The apparatus is first used as a multivariate screening system for the simultaneous determination of the narrow range between hyperoxia and hypoxia and for designing the optimal formula corresponding to the gas phase. This system can provide up to 16 types of different premixed gases into each capsule individually. The commonly used infrared CO_2 sensor and the Galvanic current O_2 sensor devices have sufficient sensitivity and undergo scheduled calibrations to maintain accuracy assurance. When a small amount of gas is infused for fine control of the gas phase, static diffusion causes an inhomogeneous gas concentration in the chamber, and the display values are often similar to those around the sensors. We therefore used pre-mixed gases and a small capsule for precise control of O_2 concentration. Pure O_2, CO_2, and N_2 gases were mixed according to their weight base molar ratios and compressed in the gas canister.. The following gases were used for cell culture experiments: 2.0% O_2, 5.0% CO_2, and 93% N_2 as an example of hypoxic culture. For purging the capsule, 5.0% CO_2 and 95% N_2 were used. The gas compositions were measured using gas chromatography, according to the pre-shipment review.

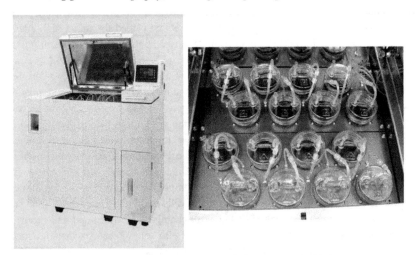

Sixteen small capsules for individual cultures were placed in a well of an aluminum dry block.

Figure 1. Individual cell culture apparatus

A   B

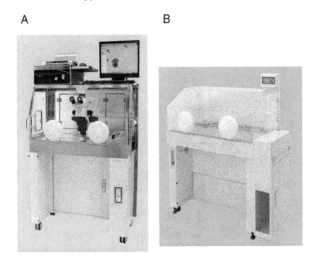

A bench top was covered with an acrylic chamber to prevent leakage of 5.0% CO_2-air. Photograph A shows the system equipped with a built-in microscope, personal computer, monitor, and printer. Photograph B shows the apparatus for general handling of cultures without microscopic observation. A minimal volume of internal space is preferable for precision control of the gas phase.

Figure 2. 5.0% CO_2-air circulation clean bench

The degree of cleanliness of air was defined by a "cleanliness class", which is specified by the number of particles of a size 0.5 μm or over in one cubic feet of air. For instance, a cleanliness class of 100 is interpreted as less than 100 particles in one cubic feet of air. The simultaneous measurements of particle size and number were performed using a light-scattering particle counter. The intake air stream was first passed through a high-intensity laser beam. As a result, the particles in the sample caused light scattering, and their numbers and intensities were detected. Room air often shows a cleanliness class of 10^6–10^5, and the aim of a conventional clean bench is to provide a low-dust environment below a cleanliness class of 10^2. The cell handling is, however, performed in ambient air, allowing temperature decrease, dissolution of O_2, and pH change by removal of CO_2. We newly developed the 5.0% CO_2-air circulation clean benches with or without a built-in microscope (Fig. 2). The bench top was covered with an acrylic chamber to prevent leakage of the ambient atmosphere, with the set-up resembling an infant incubator. Pure CO_2 was infused with the aid of a gas sensor control to maintain 5.0% CO_2-air. The bench top and the ambient temperature were kept at 37°C and 30°C–34°C by temperature control (Fig. 2). In addition, a small chamber was set on the bench top, so that if cells could not tolerate 5.0% CO_2-air for more than a few minutes, they were isolated in the chamber, and the humidified culture gas was supplied. If the bench top was contaminated with some infectious material such as body fluid, it was merely wiped off. In the newly developed system, a cover shield was placed on the bench top, and a disposable clear film was set and discarded at each operation (Fig. 3). Although the conventional clean bench filtered fresh air only once, the newly developed system circulated the enclosed 5.0% CO_2-air through a HEPA filter every 24 sec. Before starting the filtration process, the cleanliness class was found to be 10^5; however, a cleanliness class of 1 was readily achieved by repeated filtration within 5 min (Fig. 4). Furthermore, particles within the size range of 0.3 μm–0.5 μm were reduced to less than 100 within 5 min.

In the newly developed system, a bench top was shielded with a clear film for infection control.

Figure 3. Disposable cover film on bench top

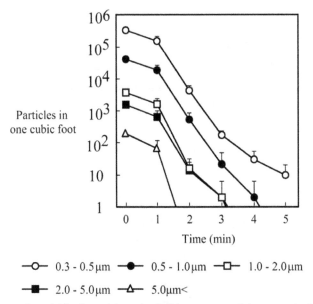

Measurements were performed at five time points on a bench. Values are represented as mean ± standard error.

Figure 4. Change in cleanliness class after starting filters

The conventional CO_2 incubator has a structural problem when it comes to achieving a stable hypoxic environment. As summarized in Table 1, whenever the door is opened, a large amount of ambient air intrudes, and the reduced CO_2 can be readily recovered by infusion of pure gas; however, it took more than 30 min to exclude O_2 by flushing with N_2. Thus, we developed a disposable small capsule to control the gas phase, especially for hypoxic tissue culture (Fig. 5). A 500-ml plastic capsule containing 220 ml of the gas buffer solution (20 mM H, 25 mM $NaHCO_3$, and 0.05% Phenol Red) was used as the CO_2 incubator. First, it was equilibrated by ventilation of the pre-mixed culture gas (10 ml/min) for at least overnight. Following gas equilibration, the pH was adjusted to 7.4 ± 0.05, and the O_2 concentration was measured using a Galvanic current O_2 sensor. Coexistence of a large amount of gas buffer solution, which serves the same function as the culture medium in terms of gas equilibration, stabilizes the physicochemical environment by functioning as a heat storage and gas pool. Inflow of air when the cap is opened should be excluded as soon as possible. To achieve this, the anoxic purging gas (5.0% CO_2 and 95% N_2) was flushed (500 ml/min) just after closing the cap. As a consequence, the oxygen level returned to 2.0% within 4 min, after which the gas supply was changed automatically to culture gas, which was infused (10 ml/min) continuously to maintain positive pressure (Fig. 6). If the gas purging process was omitted, it took 120 min until recovery, despite the inner space volume being only 280 ml (Fig. 6). This fact suggested that the void volume of the culture capsule should be minimized as much as possible, and coexistence of the gas buffer enhance the stability of the gas phase in the culture environment. In this system, gas control through a CO_2 sensor was not necessary, and we also did not need to consider the improper gas control caused by sensor deterioration. Gas equilibration of each capsule was roughly estimated by checking the color of phenol red in the gas buffer (Fig. 5), and the precision control of the culture environment was monitored by measuring the temperature and pH of the gas buffer. Although simultaneous culturing of multiple tissues is usually possible in a single CO_2 incubator, the present method allows the culturing of individual tissues in disposable capsules. The system also has additional advantages in that it allows easy and error-free identification of dishes and avoids disturbances in culture conditions when the door of the unit is opened.

A maximum of five culture dishes (6.0 cm in diameter) can be placed on the tray of the stainless stand. The gas buffer is placed at the bottom. Two tubes protruding from the cap are inlets and outlets for gas. In-line gas filters (pore size, 0.22 μm) are placed at the inlets and outlets.

Figure 5. Use of disposable capsules for individual cultures with precise control of oxygen concentration and quick recovery from disturbances in culture conditions

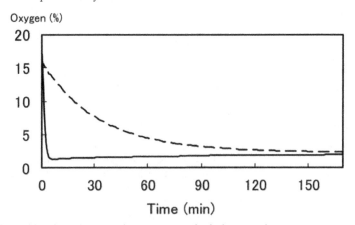

: purge with 5.0% CO_2 and 95% N_2; ------ : without purging, supply of culture gas only

Figure 6. Effect of gas purging on O_2 concentration recovery in capsule

Period of open door (sec)	0	10	20	30
CO2 (%)	5.0%	5.4%	5.4%	5.3%
Time require for recovery (min)	-	2	2	2
O2 (%)	2.1%	13.7%	18.4	18.2%
Time require for recovery (min)	-	30	34	35

The door of a conventional incubator (chamber volume: 30 L) was opened for 10, 20, and 30 sec, and the values on the display unit of the built-in device were observed. The durations required for the recovery were recorded.

Table 1. Time requirement for recovery of gas phase after door opening

The most common cultureware or vessels are sterile, disposable, and specially treated with polystyrene plastic. The cultureware includes petri dishes, multiwell plates, microtiter plates, roller bottles, and screw-cap flasks. All cultureware is equipped with lids or caps to prevent contamination from aerosol bacteria, and these culture vessels are designed to stack. Handling of culture media in conventional or in 5.0% CO_2-air circulation clean benches caused dissolution of oxygen in the media. After the lid was mounted on the culture dish or the cap of the flask was loosely closed, ambient air or 5.0% CO_2-air remained in the inner space of the cultureware. We placed an O_2 sensor on the lid of a culture dish (90 mm diameter, 10 mm height) or on the body of a culture flask (250 ml) to measure the ventilation velocity between the outer and inner spaces of the cultureware. As shown in Fig. 7, when the lid is held in the normal position and placed in the culture gas containing 2.0% O_2, it took more than 40 min for equilibration, despite the inner space volume being only about 60 ml. When the lid was held over the spacers (2 mm and 7 mm in height), the time for equilibration was again shortened to 20 min. If the cells demand a faster velocity of ventilation, a lid made out of gas-permeable materials should be used, or cultureware without lids should be used. A flask with a screw cap has a larger void volume than that of a dish, and, hence, more reliable results were obtained using a flask. The cap of the flask was opened in ambient air and closed loosely. When the flask was placed in the culture-gas environment, the ventilation velocity was found to be extremely low, and it took more than 20 h to attain equilibrium (Fig. 8-A). Moreover, the same duration was required for gas leakage, which served as a reversal process (Fig. 8-B). We examined purging of ambient air with anoxic gas in the same manner as described in Fig. 6. A needle was inserted in the cap as a gas injection port, and the flask was capped loosely (Fig. 8). The flushing (500 ml/min) of anoxic gas obviously accelerated the ventilation, and the oxygen level returned to 2.0% within 4 min (Fig. 8-C). This result suggested that the conventional use of a flask with a loose cap, which is subsequently placed in the CO_2 incubator, is unfavorable for primary and stem cells, which were intolerant to prolonged changes in pH and PO_2. An air-tight plastic vessel often suffers from gas leakage through the sealant of wide open-mouthed containers as well as due to the gas permeability of materials. After the cap was closed, the gas phase was recovered by flushing (Fig. 6), and a minimum amount of culture gas (10 ml/min) was supplied constantly (Fig. 9-A) or intermittently (Fig. 9-B) in order to maintain positive pressure. The constant supply of culture gas held PO_2 steady, whereas the intermittent supply caused narrow, wave-like changes due to gas leakage, although their margins of fluctuation were not so much different from each other. The intermittent supply saved gas consumption. The computer-assisted programmable system allowed greater flexibility to evaluate optimum environmental settings. Fig. 10 shows a time-course model of switching of the gas phase with intermittent gas supply.

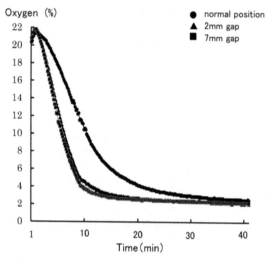

Figure 7. Effect of gap between lid and culture dish on ventilation velocity

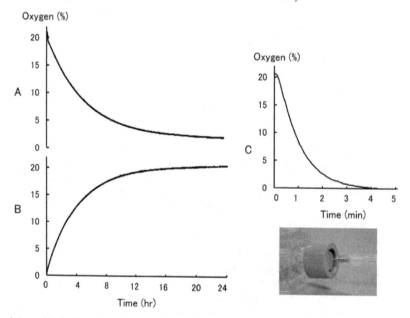

The photograph in the figure shows the experimental set-up of the injection port.

Figure 8. Ventilation velocity of loose cap flask and effects of purging with anoxic gas

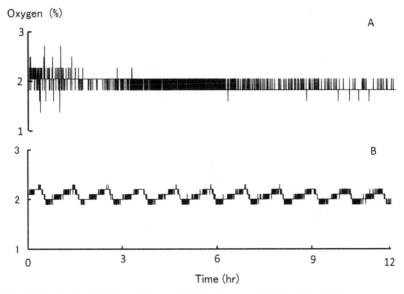

The intermittent method constituted a repetitive cycle of a 10-min gas supply followed by a 50-min pause.

Figure 9. Time-course changes in O_2 concentration during constant and intermittent gas supply

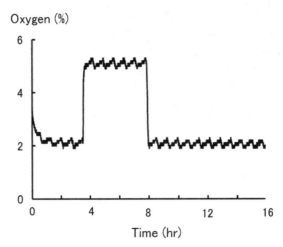

Culture gas was changed from 2.0% O_2 to 5.0% O_2 in mid-course, and then changed back to 2.0% O_2.

Figure 10. A model of computer-assisted programmable intermittent gas supply

The 16 capsules placed in the culture bath (Figs. 1 and 5) were used as a multivariate assessment system to determine the optimal formula corresponding to the narrow range between hyperoxia and hypoxia. Fig. 11 presents the model usage to optimize the combination of three parameters, namely the four premixed gases with 0% to 6.0% O_2 and the dosage of two constituents. For example, the capsule at the right edge/bottom line the combination of the constituent α, dose 4 and the constituent β, dose 5/6.0% O_2. The formula of widely used media (for example, RPMI and MEM) has been established more than half of a century ago; at that time, the multifaceted pharmacological actions and the concept of genotoxicity of some constituents had not yet been established. Amino acids are often added as supplements in the media, some of which serve as the most abundant neurotransmitters in the brain. Amino acids are responsible for almost all rapid signaling between neurons. For example, glutamate is used as a nitrogen source to promote the syntheses of proteins and nucleic acids, and it is the major excitatory neurotransmitter that is distributed in all regions of the brain ([50]. Inadequate dosing causes glutamate-induced excitotoxicity ([51]. Extracellular ATP ([52], while the decomposed species, adenosine [53], is responsible for calcium channel regulation. It is very important to evaluate whether target cells are pharmacologically sensitive to some of the constituents as well as to impurities and their degraded agents. Moreover, it is important to note that the term "sensitive" includes genotoxicity.

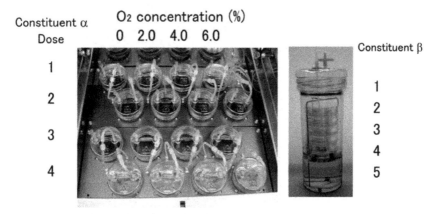

The apparatus can provide a maximum of 16 premixed gases in each capsule, and the combination of a few parameters can be determined simultaneously.

Figure 11. Multivariate assessment of environmental settings

4. Conclusion

To date faster proliferation has often been associated with the optimum culture environment, we have to investigate minutely whether this enhanced proliferation is not caused by genetic transformation or malignant changes or not. The present review dealt with "cell handling and culture under controlled oxygen concentration". The precision control of oxygen to determine the narrow range between hyperoxia and hypoxia is likely to play an important role in ensuring the safety of cell cultures, especially for primary and stem cells.

Author details

Satoru Kaneko* and Kiyoshi Takamatsu

Reproduction Center, Gynecology, Ichikawa General Hospital, Tokyo Dental College, Sugano, Ichikawa, Chiba, Japan

5. References

[1] Archibald F. Oxygen toxicity and the health and survival of eukaryote cells: A new piece is added to the puzzle. Proc Natl Acad Sci 2003; 100 (18) 10141-10143.

[2] Kazzaz JA, Xu J, Palaia TA, Mantell L, Fein AM, Horowitz S. Cellular oxygen toxicity. Oxidant injury without apoptosis. J Biol Chem 1996; 271(25) 15182-15186.

[3] Satoh T, Enokido Y, Kubo T, Yamada M, Hatanaka H. Oxygen Toxicity Induces Apoptosis in Neuronal Cells. Cellular and Molecular Neurobiol 1998; 18(6) 649-666.

[4] Yamamoto Y, Niki E, Kamiya Y, Shimasaki H. Oxidation of lipids. Oxidation of phosphatidyl cholines in homogeneous solution and in water dispersion. Biochim Biophys Acta 1984; 795(2) 332-340.

[5] Papa S. Mitochondrial oxidative phosphorylation changes in the life span. Molecular aspects and physiopathological implications. Biochim Biophys Acta 1984; 1276(1) 87-105.

[6] Bongarzone ER, Pasquini JM, Soto EF. Oxidative damage to proteins and lipids of CNS myelin produced by in vitro generated reactive oxygen species. J. Neurosci Res 1995; 41(2) 213-221.

[7] Burhans WC, Weinberger M. DNA replication stress, genome instability and aging. Nucleic Acid Res. 2007; 35(22) 7545-7556.

[8] Halliwell B. Oxidative stress in cell culture: an under-appreciated problem? FEBS Letters 2003; 540(1) 3-6.

[9] Stempien-Otero A, Karsan A, Cornejo CJ, Xiang H, Eunson T, Morrison RS, Kay M, Winn R, Harlan J. Mechanisms of Hypoxia-induced Endothelial Cell Death, role of p53 in apoptosis. *The Journal of Biological Chemistry 1999;* 274(12) 8039-8045.

[10] Steinbach J P, Wolburg H, Klumpp A, Probst H, Weller M. Hypoxia-induced cell death in human malignant glioma cells: energy deprivation promotes decoupling of mitochondrial cytochrome c release from caspase processing and necrotic cell death. Cell Death and Differentiation 2003; 10(4) 823–832.

[11] Palermo G, Jorisk K, Devroey P, Van Steirteghem AC. Pregnancies after injection of single spermatozoa into an oocyte. Lancet 1992; 340(8810) 17 -18.

[12] Green NS. Risks of Birth Defects and Other Adverse Outcomes Associated With Assisted Reproductive Technology. *Pediatrics 2004;* 114(1) 256 -259.

[13] Ludwig M. Malformation rate in fetuses and children conceived after ICSI: results of a prospective cohort study. Reproductive BioMedicine Online 2002; 5(2) 171–178.

* Corresponding Author

[14] Davies MJ, Moore VM, Willson KJ, Van Essen P, Kevin Priest K, Scott H, Mgmt B., Eric A. Haan EA, Chan A. Reproductive Technologies and the Risk of Birth Defects. N Engl J Med 2012; 366(19) :1803-1813.

[15] Marchetti F, Essers J, Kanaar R, Wyrobek AJ. Disruption of maternal DNA repair increases sperm-derived chromosomal aberrations. Proc Natl Acad Sci U S A 2007; 104(45) 17725–17729.

[16] Evenson DP, Jost LK, Marshall D, Zinaman MJ, Clegg E, Purvis K, de Angelis P, Claussen OP. Utility of the sperm chromatin structure assay as a diagnostic and prognostic tool in the human fertility clinic. Hum Reprod 1999; 14(4) 1039-1049.

[17] Zini A, Boman JM, Belzile E, Ciampi A. (2008) Sperm DNA damage is associated with an increased risk of pregnancy loss after IVF and ICSI: systematic review and meta-analysis. Hum Reprod 2008; 23(12) 2663-2668.

[18] Aitken RJ, De Iuliis GN. Origins and consequences of DNA damage in male germ cells. Reprod Biomed Online 2007; 14(6) 727-733.

[19] Aitken RJ, De Iuliis GN. Value of DNA integrity assays for fertility evaluation. Soc Reprod Fertil Suppl 2007; 65(1) 81-92.

[20] Simon L, Brunborg G, Stevenson M, Lutton D, McManus J, Lutton D, McManus J, Lewis SEM., Clinical significance of sperm DNA damage in assisted reproduction outcome. *Hum Reprod* 2010; 25(7) 1594-1608.

[21] Irvine DS, Twigg JP, Gordon EL, Fulton N, Milne PA, Aitken RJ. DNA integrity in human spermatozoa: relationships with semen quality. J Androl 2000; 21(1) 33-44.

[22] Hindryckx A, Peeraer K, Debrock S, Legius E, de Zegher F, Francois I, Vanderschueren D, Demyttenaere K, Rijkers A, D'Hooghe T. Has the prevalence of congenital abnormalities after intracytoplasmic sperm injection increased? The Leuven data 1994–2000 and a review of the literature. Gynecol Obstet Invest 2010; 70(1), 11–22.

[23] Ludwig M. Malformation rate in fetuses and children conceived after ICSI: results of a prospective cohort study. Reproductive BioMedicine Online 2002; 5(2) 171–178.

[24] Wen SW, Leader A, White RR, Le'veille' MC, Wilkie V, Zhou J & Walker MC. A comprehensive assessment of outcomes in pregnancies conceived by in vitro fertilization / intracytoplasmic sperm injection. Eur J Obstet Gynecol Reprod Biol 2010; 150(2), 160–165.

[25] Alukal JP, Lamb DJ. Intracytoplasmic sperm injection (ICSI) - what are the risks? Urol Clin North Am 2008; 35(2), 277–288.

[26] Mastroianni L, Jones R. Oxygen tension within the rabbit fallopian tube. *J Reprod Fertil* 1965; 9(1) 99-102

[27] Sugawara H, Iwamoto K, Bundo M, Ueda J, Ishigooka J, Kato T. Comprehensive DNA methylation analysis of human peripheral blood leukocytes and lymphoblastoid cell lines. Epigenetics 2011 6(4) 508-515.

[28] Grafodatskaya D, Choufani S, Ferreira JC, Butcher DT, Lou Y, Zhao C, Scherer SW, Weksberg R. EBV transformation and cell culturing destabilizes DNA methylation in human lymphoblastoid cell lines. Genomics 2010; 95(2) 73-83.

[29] Lister R, Pelizzola M, KidaYS, Hawkins RD, Nery JR, Hon G, Antosiewicz-Bourget J, O'Malley R, Castanon R, Klugman S, Downes M, Yu R, Stewart R, Ren B, Thomson JA,

Evans RM, Ecker JR. Hotspots of aberrant epigenomic reprogramming in human induced pluripotent stem cells. Nature 2011; 471(7336) 68-73.

[30] Tanasijevic B, Dai B, Ezashi T, Livingston K, Roberts RM, Rasmussen TP. Progressive accumulation of epigenetic heterogeneity during human ES cell culture. Epigenetics 2009; 4(5) 330-338.

[31] Tanasijervic B, Dai B, Ezashi T, Livingston K, Roberts RM, Rasmussen TP. Progressive accumulation of epigenetic heterogeneity during human ES cell culture. Epigenetrics 2009; 4(5) 330-338.

[32] Bavister BD. Culture of preimplantation embryos: facts and artifacts. Hum. Reprod. Update 1995; 1 (2): 91-148.

[33] Palermo GD, Neri QV, Takeuchi T, Squires J, Moy F, Rosenwaks Z. Genetic and epigenetic characteristics of ICSI children. Reprod Biomed Online 2008; 17(6):820-33.

[34] Kohda T, Ogonuki N, Inoue K, Furuse T, Kaneda H, Suzuki T, Kaneko-Ishino T, Wakayama T, Wakana S, Ogura A, Ishino F. Intracytoplasmic sperm injection induces transcriptome perturbation without any transgenerational effect. Biochem Biophys Res Commun. 2011;410(2):282-288.

[35] Quinn P. Harlow GM. Effect of oxygen on the development of preimplantation mouse embryos in vitro. J Exp Zool 1978; 206(1) 73–80.

[36] Pabon J E, Findley WE, Gibbons WE. The toxic effects of short exposures to the atmospheric oxygen concentration on early mouse embryonic development. Fertil Steril 1989; 51(5) 896–900.

[37] Umaoka ., Noda Y, Narimoto K, Mori T. Effects of oxygen toxicity on early development of mouse embryos. Mol. Reprod. Dev. 1992; 31(1) 28–33.

[38] Umaoka Y, Noda Y, T. Nakayoma T, Narimoto K, Mori T, Iritani A. Development of hamster one-cell embryos recovered under different conditions to the blastocyst stage. Theriogenology 1993; 39(2):485–498.

[39] Thompson JGE, Simpson AC, Pugh PA, Donnelly PE, Tervit HR. Effect of oxygen concentration on in-vitro development of preimplantation sheep and cattle embryos. J. Reprod. Fertil 1990; 89(2) 573–578.

[40] Nakao H, Nakatsuji N. Effects of co-culture, medium components and gas phase on in vitro culture of in vitro matured and in vitro fertilized bovine embryos. Theriogenology 1990; 33(3) 591 - 600.

[41] Nasr-Esfahani MH, Winston NJ, Johnson MH. Effects of glucose, glutamine, ethylenediaminetetraacetic acid and oxygen tension on the concentration of reactive oxygen species and on development of the mouse preimplantation embryo in vitro. J Reprod Fertil 1992; 96(1) 219–231.

[42] Burton GJ, Hempstock J, Jauniaux E. Oxygen, early embryonic metabolism and free radical-mediated embryopathies. Reproductive BioMedicine Online 2003; 6(1) 84–96.

[43] Gwatkin RBL, Haidri AA. Requirements for the maturation of hamster oocytes in vitro. Exp. Cell Res. 1973;76(1):1–7.

[44] Mehta TS, Kiessling AA. Developmental potential of mouse embryos conceived in vitro and cultured in ethylenediaminetetraacetic acid with or without amino acids or serum. Biol Reprod 1990; 43(4): 600–606.

[45] Halliwell B, Gutteridge JMC. Oxygen toxicity, oxygen radicals transition metals and disease. Biochem. J 1984; 219(1) 1–14.

[46] Catt JW, Henman M. Toxic effects of oxygen on human embryo development. Human Reproduction 2000; 15(Suppl. 2) 199-206.

[47] Rytting E, Audus KL. Effects of low oxygen levels on the expression and function of transporter OCTN2 in BeWo cells. J Pharm Pharmacol 2007; 59(8) 1095-1102.

[48] Hirao M, Hashimoto J, Yamasaki N, Ando W, Tsuboi H, Myoui A, Yoshikawa H. Oxygen tension is an important mediator of the transformation of osteoblasts to osteocytes. J. Bone Miner. Metab. 2007; 25(5) 266-276.

[49] Kaneko S, Takamatsu K, Yoshida J, Miyaji K, Ishikawa H, Kawamata T, Shinozaki N. Individual tissue culture system in a disposable capsule with hypoxic atmosphere. Ann Cancer Res Therap 2008; 16(1). 8-11.

[50] Meldrum BS. "Glutamate as a neurotransmitter in the brain: Review of physiology and pathology". Journal of nutrition 2000; 130(4S Suppl) 1007S-1015S.

[51] Choi DW, Maulucci-Gedde M, Kriegstein AR. Glutamine neurotoxicity in cortical cell culture. J. Neurosci, 1987; 7(2) 357-368.

[52] Liu QY, Rosenberg RL, Stimulation of cardiac L-type calcium channels by extracellular ATP. *Am J Physiol Cell Physiol* 2001; 280(5) C1107-C1113.

[53] Stella Jr. SL, Bryson EJ, Thoreson WB. A$_2$ Adenosine Receptors Inhibit Calcium Influx Through L-Type Calcium Channels in Rod Photoreceptors of the Salamander Retina. *AJP - JN Physiol* 2002; 87(1) 351-360.

Chapter 4

In vitro, Tissue-Based Models as a Replacement for Animal Models in Testing of Drugs at the Preclinical Stages

Zhanqiu Yang and Hai-Rong Xiong

Additional information is available at the end of the chapter

1. Introduction

As early as 1950s, researchers began to apply *in vitro* culture technology in testing cytotoxicity of multiple drugs on different cells, which could basically determine the cytoxic dose range of these drugs. Regarding to the continuous development of novel chemotherapy drugs and large growth of chemical compounds closely related to human practice and life (including pharmaceutical, cosmetics, food additives, pesticide, industrial chemical, etc), there is great need to explore a convenient and effective way for selection, pre-clinical evaluation or pre-production safety assessment.

In vitro, tissue-based models are common and widely used for screening and ranking chemicals, especially in testing of drugs at the preclinical stages. The toxic effects include general cytotocixity, genotoxicity, mutagenesis and carcinogenesis. Cell-based assays are currently considered central to toxicity testing, biomaterial testing, and environmental material exposure testing. Nearly all of the assays could be adapted to other application for bioactivity test. For instance, the established cell lines have been successfully employed in a number of fields of medical research. Especially, many aspects of modern virology have been developed using animal cells in culture.

There are several strategies for using in vitro, tissue-based models in testing of drugs at the preclinical stages. One such strategy is to refine the choice of cells and end points of one method. For instance, human corneal cells are now used to screen for local eye toxicity of chemicals, with a method employing sophisticated end points. Another strategy is to use batteries of tests with different cell types, to cover most aspects of basic cell functions. A third strategy is to do more basic research into fundamental mechanisms of toxicity or bioactivity. When such mechanisms have been clarified, rational in vitro models could be set up. By contrast, a fourth approach ignores whether the toxic mechanisms screened for is known or not. As long as the end point of the test correlates well with in vivo toxicity, the test may be used.

2. Cultures

The possibility of using cell or tissue cultures as suitable material for testing agents in pharmacology has often been suggested. The culture methods include organ culture, sphere culture, suspension culture, clone culture on soft agar and monolayer cell culture, etc.

Organ culture provides drug evaluation a model more close to in vivo situation, but this model are not suitable for efficacy quantitative experiment because of the size differentiation of organ implants and heterogeneity of the cells within the explants between repeated experiments. Sphere culture of the tumor cells is similar to tumor nodules in vivo and can be used to study the influence of three-dimensional relationship on drug sensitivity. Suspension culture is beneficial to furthest prevent the growth of "cell pollution source"- connective tissue cells (i.e. fibroblast), which is widely applied to chemosensitivity study and convenient to perform radionuclide analysis. Clone culture on soft agar is fit for cell with high capacity of self-renewal (i.e. tumor cell), other than cells with limited proliferation ability. Monolayer cell culture is best for cytotoxicity test of cell lines and chemosensitivity of various tumor biopsy materials. This test can implement automatic operation due to small cell amount requirement and its convenience and flexibility in drug treatment, reply and pharmacodynamic quantitative study.

Generally, a primary cell line possesses many characteristics of the original cells, such as similar chromosomal numbers and the specialized biochemical properties of the parent tissue, e.g. in the case of liver cells the ability to secrete albumin. On the contrary established cell lines invariably have different chromosome numbers and lose a number of specialized biochemical properties of the parent tissues. This latter fact imposes some restrictions on the utilization of established cultures in the design of tests of toxicity. However, established cell lines have been successfully employed in a number of fields of medical research. Especially, many aspects of modern virology have been developed using animal cells in culture.

3. Method used

3.1. Cytotoxicity

Cytotoxicity is considered primarily as the potential of a compound to induce cell death. Most in vitro cytotoxicity tests measure necrosis. However, an equally important mechanism of cell death is apoptosis, which requires different methods for its evaluation. Moreover, detailed studies on dose and time dependence of toxic effects to cells, together with the observation of effects on the cell cycle and their reversibility, can provide valuable information about mechanisms and type of toxicity, including necrosis, apoptosis or other events.

In vitro cytotoxicity tests are useful and necessary to define basal cytotoxicity, for example the intrinsic ability of a compound to cause cell death as a consequence of damage to basic cellular functions. Cytotoxicity tests are also necessary to define the concentration range for further and more detailed in vitro testing to provide meaningful information on parameters such as genotoxicity, induction of mutations or programmed cell death. By establishing the dose at which 50% of the cells are affected (i.e. TC_{50}), it is possible to compare quantitatitively responses of single compound in different systems or of several compound in individual systems.

The endpoint/parameters used in cellular toxicity testing including:

- Membrane permeability changes-dye exclusion(trypan blue);the release of intracellular enzymens like lactate dehydrogenase; preloaded ^{51}Cr; nucleoside release; uridine uptake;vital dye uptake, etc.
- Reduced mitochondrial function
- Changes in cell morphology
- Changes in cell replication
- Apoptosis evaluation-changes in morphology; membrane rearrangements; DNA fragmentation; caspase activation; cytochrome c release from mitochondria, etc

3.2. Cellular response and functional response regarding to general toxicity(protein/gene expression)

The basic methodology of general toxicity has changed little during past decades. Toxicity in laboratory animals has been evaluated by mainly using clinical chemistry, hematological and histological parameters as indicators of organ damage. The effect of a toxic chemical on a biological system in most cases is fundamentally reflected, at the cellular level, by its influence on gene expression. Consequently, measurement of the transcription (mRNA) and translation (protein) products of gene expression can explore valuable information about the potential toxicity of chemicals before the development of a toxic/pathological response.

The rapid progress in genomic (DNA sequence), transcriptomic (gene expression) and proteomic (the study of proteins expressed by a genome, tissue or cell) technologies, in combination with the ever-increasing power of bioinformatics, creates a unique opportunity to form the basis of improved hazard identification for more predictive safety evaluation.

For example, currently available methods for the study of gene expression at the transcript level include:

- Hybrdizization-based techniques: Northern blotting;S1-mapping/RNase protection; Differential plaque hybridization;
- PCR-based techniques: Subtraction cloning; DNA microarrays; Differential display; RDA(representational difference analysis); Quantitative (real time) PCR;
- Sequence-based techniques:ESTs(expressed sequence tags); SAGE(serial analysis of gene expression); MPSS(massively parallel signature sequencing); DNA-sequencing chip; Mass-spectrometry sequencing

Two-dimensional gel electrophoresis is a highly sensitive means of screening for toxicity and probing toxic mechanisms, which combines separation of proteins by isoelectric focusing(IEF) in the first dimension followed by sodium dodecyl sulphate-polyacrylamide gel electrophoresis(SDS-PAGE) based on molecular weight in the second dimension.

By comparing gene/proteins expressed following exposure of a biological test system to a chemical with those present under untreated conditions, it is possible to identify changes in biochemical pathways via observed alteration in sets of gene/proteins that may be related to the tocicity, which provide the means to profile expression of thousands of messenger RNAs or proteins.

Over the last few decades, a large amount of research has resulted in an explosion of information regarding mechanisms of toxicity and new tools to study the biological responses to toxic stress. Some of the key cellular responses to toxicant exposure, which could potentially be used as early markers of toxicity, include the following:

- Responses following exposure to toxicants that form reactive electrophiles (e.g. oxidative stress) such as loss of glutathione (GSH), increased production and sensitivity to reactive oxygen species (ROS), increase in cellular calcium, lipid peroxidation, loss of ATP and mitochondrial/endoplasmic reticulum (ER) specific events.
- The cellular response to stress, including an increase in synthesis of the heat shock (Hsp) family of proteins, induction of the stress-activated protein kinases (SAPKs) and glucoseregulated proteins (Grps).
- Changes in the levels of key enzymes, such as the phase I and phase II metabolising enzymes involved in the detoxification of toxic chemicals.
- Induction of the metal-binding proteins, metallothioneins (MTs).
- Perturbations to cellular membranes, gap junctions and intercellular communication inhibition (involving the connexins Cx43, Cx32 and Cx26).
- Induction of cell proliferation (for which suitable markers could include TNF-α, TNF-β, plasminogen activator inhibitor-2 (PAI-2), the tumour proliferative marker Ki-67 antigen and proliferating cell nuclear antigen (PCNA).

3.3. Toxicokinetic study

Toxicokinetic modelling describes the absorption, distribution, metabolism and elimination of xenobiotics as a function of dose and time within an organism. Toxicokinetic models can be divided into two main classes: data-based compartmental models and physiologically-based compartmental models. In vitro approaches can be used to obtain useful information in this area. One representative model, the physiologically-based toxicokinetic (PB-TK) model, can be obtained by studies in vitro, including tissue–blood partition coefficients, the kinetics of any active transport processes, and the kinetics of metabolism by the liver and any other organ capable of biotransforming the compound (e.g. the lung). The output of toxicokinetic models is the prediction of concentration/time courses in different tissues. This information can be combined with the basal cytotoxicity data to make a prediction of the acute systemic toxicity of the chemical.

3.4. Specific toxicity

3.4.1. Genotoxicity

In vitro system can also be used to determine genotoxicity for identifying potential carcinogens, which includes three levels of mutation, namely *gene, chromosome and genomic* mutations. It is acknowledged that the generally accepted objectives of genotoxicity testing of chemicals are: (a) identification of germ cell mutagens, because of their possible involvement in the etiology of human heritable genetic defects; (b) identification of somatic cell mutagens, because of their involvement in neoplastic transformation.

As far as test methods are concerned, it is recommended that OECD protocols be used, which updated six previous guidelines and introduced a new one in 1997. These guidelines provide guidance for the conduct of in vitro screening tests (e.g. gene mutations in bacteria and in mammalian cells, chromosomal aberrations in vitro) as well as for the in vivo assays (e.g. micronuclei and chromosomal aberrations in rodent bone marrow, rat liver unscheduled DNA synthesis, chromosomal aberrations in spermatogonia).

A number of useful techniques have been developed with which it is also possible to determine the genetic toxicology. For example, a protocol for in vitro micronucleus test is currently being evaluated for inclusion, which might be considered in test batteries as an alternative to in vitro chromosomal aberration assay. By using centromeric specific probes based on fluorescence in situ hybridization (FISH), chromosome loss and non-disjunction specific probes allow rapid scoring of aneuploidy in a variety of cell types, including human cells. mammalian cell assays should be routinely performed according to the standard updated protocols for the detection of either gene mutation (at *tk*, HPRT, or other loci), or structuralchromosomal aberrations by metaphase analysis.

On the other hand, non-genotoxic carcinogenicity of the compound is able to be evaluated through the following common mechanism:

- Persistent cytotoxicity accompanied by proliferative regeneration.
- Chronic inflammation
- Hormones
- Ligands for xenobiotic receptors
- DNA methylation

Furthermore, there are also some potential short-term tests designed for non-genotoxic carcinogens. They are:

- Detection of mitogenesis
- The application of gene arrays and other approaches to transcription profiling
- In vitro cell transformation assays
- Transgenic cell systems
- Cytosine methylation
- Quantitative structure-activity relationships(QSAR) and other computational approaches

Overall, no single system will be will be adequate to detect all non-genotoxic carcinogens or even a large number of them. However, by focusing on those mechanisms that appear to be of relevance to humans, it may be possible to identify key toxicological responses, which provide a clear indication of carcinogenic potential.

3.4.2. Developmental toxicity

Reproductive toxicology embraces studies on male and female fertility and on developmental toxicity, with special emphasis on embryotoxicity and teratogenicity. Over the past 20 years, more than 30 different culture systems have been proposed as tests for developmental toxicity. The culture systems fall into the following categories:

- Tests on non-vertebrate species, including Hydra, slime moulds, brine shrimps and Drosophila.
- Tests on lower vertebrate embryos or embryonic cell aggregates, including fish, amphibians and birds.
- Tests on whole mammalian embryos.
- Tests on micromass cultures from mammalian embryos (limb buds, midbrain).
- Tests on embryonic stem cells or embryonic stem cell lines.
- Tests on other mammalian cell lines (e.g. human embryonic palate mesenchymal cells, mouse ovarian tumour cells, neuroblastoma cells, teratocarcinoma cells).

3.4.3. Cell lines and embryonic stem cells

A number of established cell lines have been used for screening purposes in 1980s. These include: human embryonic palate mesenchymal cells, mouse ovarian tumour cells and neuroblastoma cells. The results show a high number of false positives.

However, the use of omnipotent embryonic stem cell lines shows more promising results. For instance, blastocyst totipotent embryonic stem cells (ES) can be cultured under conditions in which the cells form several types of differentiated cells, such as muscle cells or haematopoietic cells. These culture systems can be used to determine the two essential features of embryotoxicity: inhibition of differentiation and/or a higher sensitivity to cytotoxic effects in embryonic cells than adult tissues. Results of such a test were comparable to the outcome of an embryotoxicity test with rat whole embryo cultures. The use of ES cells in the production of transgenic cells with targeted mutations and reporter constructs should enable the development of tests with simplified endpoints, which can be used in robotised assay systems. What's more, new developments in which multipotent (or even totipotent) stem cells can be isolated from adult tissues are very promising. For example, nervous tissue stem cells can give raise to haematopoietic stem cells and vice versa.

Besides, testing systems using aggregate and micromass cultures, embryos of lower order species, avian and mammaliam whole embryo culture for detecting the developmental toxicity have been described and widely used. They allow the detection of dysmorphogenesis in many organs and the comparison of specific dysmorphogenic effects with general adverse effects on growth and differentiation. In addition, they enable the potencies of structurally related compounds to be ranked. Limitations of these systems are related to the fact that they are relatively complex, cover only a part of organogenesis, require high technical skills and they also can be costly.

In conclusion, in vitro culture detection system is more suitable for anti-cancer drug and teratogenic, carcinogenic, and mutagenic chemicals. Main application area include that 1) Identify potential active compounds; 2) Study the mechanism of chemical toxicity; 3) Predict effective toxicity drug possibly used in treatment for cancer patients; 4) Screen effective component and its active range from a variety of compounds; 5) Determine the types of effector cells; 6) Confirm toxicity range; 7) Investigate relationship between the drug concentration and exposure time.

Although the drug efficacy, adverse reaction and safety evaluation of need certain animal experiment, the *in vitro* culture technology has already become common tool for testing cytotoxicity and efficacy globally. The advantages are that 1) established cell lines provide cell sources with uniform or similar genetic background for in vitro test, which make the drug selection more stable, convenient and economic.2) This technique can precisely control the object, time and dose of drug action as well as the cell growth conditions. For example, clinical medicine study can choose human cells. Tissue-specific drugs need to select targeted and relevant cells. Researchers employ tumor cell lines to evaluate anti-cancer drugs by observing the cell reaction to drug from cytological view. Therefore, this in vitro detection avoid the drawbacks of in vivo experiment about lack of specific effector cell, drug metabolism reaction, different drug resistance between species and individuals. 3) It's easier to distinguish the direct drug effect and indirect internal effect by adding drugs or other chemicals directly to cell culture system, or even injecting drug into the cell. 4) Animal protection moral appeal reduction of animal experiment internationally. Thus the application of in vitro technology in drug test becomes more and more widespread.

Definitely, determining the action of different drugs on the cell in vitro under various dosages is not completely suitable for in vivo status. The disadvantages of in vitro test are 1) in vivo growing environment is quite complicated compared to the simple in vitro culture condition. The in vivo experiments possess not only comprehensive functional regulating system, but also the metabolism, modification (such as increase or decrease of pharmaceutical biochemistry through the liver, kidneys and other organs), immunological effect. 2) Cytotoxicity and activity test using in vitro system is suitable for monomer drug other than compound medicines and Chinese patent medicine. 3) It is easier to test water-soluble drug compared to the water insoluble drug, which may need suitable solvent to dissolve and control group in experimental design to exclude the effect of the solvent itself to the cell. 4) In vitro system is mainly applicable to acute toxicity research, thus analysis of chronic effect requires improving the culture system.

Author details

Zhan-Qiu Yang and Hai-Rong Xiong
School of Basic Medical Sciences, Wuhan University, The People's Republic of China

4. References

Aoki, Y., Lipsky, M.M., Fowler, B.A., 1990. Toxicology and Applied Pharmacology.

Hartung, T., Gstraunthaler, G., 2000. The standardisation of cell culture practices. In: Balls, M., Van Zeller, A.-M., Halder, M.E. (Eds.), Progress in the Reduction, Refinement and Replacement of Animal Experiments. Elsevier, Amsterdam, pp. 1655–1658.

Kramer, P.J., 1998. Genetic toxicology. Journal of Pharmacy and Pharmacology 50, 395–405.

Ekwall B. Screening of toxic compounds in mammalian cell cultures. Ann N Y Acad Sci. 1983;407:64-77.

Rees KR. Cells in culture of toxicity testing: a review.J R Soc Med. 1980 Apr;73(4):261-4.

Chapter 5

Isolation of Breast Cancer Stem Cells by Single-Cell Sorting

Phuc Van Pham, Binh Thanh Vu, Nhan Lu Chinh Phan, Thuy Thanh Duong, Tue Gia Vuong, Giang Do Thuy Nguyen, Thiep Van Tran, Dung Xuan Pham, Minh Hoang Le and Ngoc Kim Phan

Additional information is available at the end of the chapter

1. Introduction

Breast cancer is the most common cancer in women, with more than 1,000,000 new cases and more than 410,000 deaths each year [38]; [39]. At present, breast cancer is mainly treated by surgical therapy as well as cytotoxic, hormonal and immunotherapeutic agents. These methods achieve response rates ranging from 60 to 80% for primary breast cancers and about 50% of metastases [22]; [24]. However, up to 20 to 70% of patients relapse within 5 years [10].

The reason for recurrence is the existence of cancer stem cells in malignant tumors such brain, prostate, pancreatic, liver, colon, head and neck, lung and skin tumors [3]; [7]; [14]; [15]; [21]; [32]; [49]; [51]. Breast cancer stem cells (BCSCs) were first detected by Al-Hajj et al. (2003) that showed cells expressing CD44 protein and weakly or not expressing CD24 protein could establish new tumors in xeno-grafted mice. Using these markers, researchers isolated BCSCs from primary [41]; [47] and established breast cancer cell lines [16]. Another technique used is cell culture in serum-free medium to form mammospheres. Mammospheres exhibit many stem cell-like properties such as differentiation into all three mammary epithelial lineages [11]; [12]. These BCSCs have been demonstrated to cause treatment resistance and relapse. Thus, BCSC-targeting therapy is considered a promising therapy for treating breast cancer.

Recently, BCSC-targeting therapies have been researched by various groups worldwide. Strategies include targeting the self-renewal of BCSCs [30]; [31], indirectly targeting the microenvironment [29]; [50]; [31] and directly killing BCSCs by chemical agents that induce differentiation [25]; [19]; [42]; [43], immunotherapy [4]; [5]; [40] and oncolytic viruses [26]; [34]. In all strategies, isolation of BCSCs is an important step to recover starting materials for all subsequent steps. Thus, isolation of BCSCs is a pivotal step for successful outcomes. Almost all studies have focused more on treatment strategies than isolation of BCSCs. Indeed, to date, there are only three methods used to identify and isolate BCSCs, namely fluorescence-activated cell sorting (FACS) based on BCSC markers such as CD44, CD24 and CD133 [2]; [52]; [46]; [41], identification of the side population (SP) that effluxes Hoechst 33342 [13]; [8]; [28] and mammosphere formation [44]; [54]. All these methods possess some limitations.

The first limitation is the resulting heterogenous population of BCSCs. Using these techniques, the BCSC population contains phenotypes with differences in CD44 and CD24 expression levels. These differences reflect variations in some cellular behaviors. BCSCs isolated by SP sorting or mammosphere culture may contain a small population that do not exhibit the CD44+CD24- phenotype. Thus, in this study, we attempted to establish a new method to isolate a homogenous population from malignant breast tumors.

Our study is based on the cell cloning technique that is applied to select hybridomas for monoclonal antibody production. Using a cell sorter with the index sorting function, we aim establish a new protocol that can isolate and establish BCSC clones at a high efficiency.

2. Materials and methods

2.1. Primary culture

Primary culture of breast cancer cells from malignant breast tumors was carried out as described elsewhere [41]; [42]. Tumor biopsies were obtained from consenting patients at the Oncology Hospital in Ho Chi Minh city, Vietnam, and then transferred to the laboratory on ice. All samples were kept in phosphate-buffered saline (PBS) containing 1× antibiotics and an antimycotic (Sigma-Aldrich, St Louis, MO). Tumors were homogenized into small fragments (approximately 1–2 mm3) using scissors. These samples were seeded in 35-mm culture dishes (Nunc, Roskilde, Denmark) in M171 medium (Invitrogen, Carlsbad, CA) containing mammary epithelial growth supplement (MEGS) (Invitrogen, Carlsbad, CA), and incubated at 37°C with 5% CO_2. Five patients participated in this study.

2.2. Single-cell sorting

Primary cells were detached by 0.25% trypsin/EDTA. The cell suspension was washed twice with PBS to eliminate trypsin. The cell pellet was resuspended in sorting buffer (PBS containing 0.2 mM EDTA and 1 mg/mL bovine serum albumin (BSA) at 1×10^6 cells/ml. Single-cell sorting was performed on a BD FACSJazz (BD Bioscience, Franklin Lakes, NJ) using the index sorting function. One cell was sorted into one well of a 96-well plate that contained M171 medium with MEGS. One sample was sorted into 2880 wells in 30 plates. After sorting, all wells were checked for a single cell/well under an inverted microscope.

2.3. Single cell-based culture and selection of mesenchymal-like cells

After single-cell sorting, cells were cultured in M171 medium containing MEGS and incubated at 37°C with 5% CO_2. Half medium volumes were exchanged every 3 days for 30 days. Then, only wells that contained cell colonies were used to select mesenchymal-like cell clones by replacing M171 medium with DMEM/F12 supplemented 10% fetal bovine serum (FBS) for 2 days. In this medium, all epithelial-like cells did not survive. Surviving mesenchymal-like cell clones were continuously subcultured for three to five passages. Cell clones that rapidly underwent an epithelial-mesenchymal transition (EMT) were considered as BCSC candidates. These cells were used to analyze some characteristics of BCSCs in subsequent experiments.

2.4. CD44+CD24-/dim-based cell sorting

BCSCs were isolated from primary cultures based on CD44+CD24- expression by FACS as described elsewhere [42]. Briefly, 1 ml cell suspensions in PBS (1×10^7 cells) were double stained with 20 μl anti-CD44-FITC and 20 μl anti-CD24-PE. Samples were incubated in the dark at room temperature for 45 min. The CD44+CD24-/dim cell population was identified by the software controlling the BD FACSJazz. Cells were sorted into 2-ml tubes containing 1 ml culture medium (DMEM/F12 supplemented with 10% FBS and a 1× antibiotic-mycotic (Sigma-Aldrich, St Louis, MO).

2.5. Rhodamine 123 efflux and SP analysis

Cells were stained with 0.1 μg/mL rhodamine (Sigma-Aldrich) at 37°C for 30 min, and then washed twice with 2% FBS in PBS. Two filters (FL1 and FL3) were used to detect rhodamine 123. Cells incubated with 50 μM verapamil and 0.1 μg/ml rhodamine 123 for 30 min were used as a positive control.

2.6. Immunophenotyping

Cell markers were analyzed following a previously published protocol [43]. Briefly, cells were washed twice in PBS containing 1% BSA (Sigma-Aldrich). Cells were stained for 30 min at 4°C with anti-CD13-FITC, anti-CD90-PE and anti-CD133-PerCP monoclonal antibodies (BD Biosciences). Stained cells were analyzed by a BD FACSCalibur (BD Biosciences) flow cytometer. Isotype controls were used in all analyses.

2.7. Mammosphere culture

Cells were detached with 0.25% trypsin/EDTA and resuspended in serum-free DMEM/F12 (1:1; GeneWorld, Ho Chi Minh, VN) containing 15 ng/ml basic fibroblast growth factor, 20 ng/ml epidermal growth factor (EGF), 2 mM/l L-glutamine, 4 U/l insulin growth factor (Sigma-Aldrich) and B27 supplement (1:50; Invitrogen). Cells were cultured at 37°C with 5% CO2.

2.8. Cell cycle analysis

Cells were washed twice in PBS and fixed in cold 70% ethanol for at least 3 h at 4°C. Then, cells were washed twice in PBS and stained with 1 ml PI (20 μg/ml). Fifty microliters of RNase A (10 μg/ml) was added to samples, followed by incubation for 3 h at 4°C. Stained cells were analyzed by flow cytometry using CellQuest Pro software (BD Biosciences).

2.9. Doxorubicin resistance analysis

BCSCs were cultured to 10^4 cells/well in a 24-well plate (Nunc, Roskilde, Denmark), in DMEMF12/10% FBS. After 24 h culture to confluence, cells were treated with 0, 1 and 3 µg/ml doxorubicin (Sigma-Aldrich). Doxorubicin resistance was analyzed by apoptosis using annexin-V-FITC and PI on a FACSCalibur.

2.10. *In vivo* tumorigenesis

Non-obese diabetic (NOD)/SCID mice (5–6-weeks-old) (NOD.CB17-Prkdcscid/J; Charles River Laboratories) were used in this study. All mice were housed in clean cages and maintained according to institutional guidelines on animal welfare. Mice were subcutaneously injected with 1×10^6, 10^5, 10^4 and 10^3 cells (n=3, each dose). Mice were followed up for 1 month to detect tumors.

2.11. Statistical analysis

All experiments were performed in triplicate. A value of $P \leq 0.05$ was considered significant. Data were analyzed using Statgraphics v 7.0 software (Statgraphics Graphics System, Warrenton, VA).

3. Results

3.1. Primary culture

The study was carried out to primary culture five tumors from five patients. There were 3/5 samples that outgrew cells (Fig. 1A). These cells from the three samples were propagated until 80% confluence. In almost all samples, epithelial-like cells appeared before mesenchymal-like cells, which spread out from tumor fragments from day 5. Mesenchymal-like cells usually appeared at day 20. Then, cells proliferated rapidly and formed colonies. At this time, two cell shapes were mostly observed in the primary culture (Fig. 1B). These were epithelial-like cells with a bean shape and large nucleus, and mesenchymal-like cells with a small nucleus and elongated shape. Cells were subcultured once to expand enough cells for further experiments. The primary cells from these samples were used in both single-cell sorting and CD44⁺CD24⁻/dim-based cell sorting.

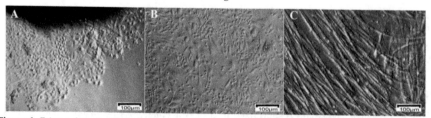

Figure 1. Primary breast cells derived from malignant tumors. (A) Primary cells began to migrate from tumors. (B) Primary cells rapidly proliferated with two main shapes indicating stromal and epithelial-like cells. (C) Cell populations were subcultured for a second passage and cultured in DMEM/F12 supplemented with 10% FBS.

3.2. Single cell-based culture, cell selection and EMT

Primary cells were individually sorted into the wells of 96-well plates. A total of 2880 wells were used for one primary cell population. Single cells were cultured in M171 medium containing MEGS for 2 weeks. There were 14.67±5.13 colonies formed per 96-well plate. Single cells in other wells did not proliferate or died (n=3). Similarly to primary culture, there were two kinds of cell clones. One kind of cell clone was epithelial-like, and the other was mesenchymal-like. To enrich mesenchymal-like cells and eliminate epithelial-like cells, we changed the medium from M171 medium containing MEGS to DMEM/F12 supplemented with FBS that was suitable for mesenchymal-like cells. After all epithelial-like cell clones died in 24 h, mesenchymal-like cell clones continued to expand to 70–80% confluence and were then subcultured. From one sample, 6.33±3.06 mesenchymal-like cell clones were derived per 96-well plate (Fig. 1C). These cell clones were subcultured continuously for three to five passages to identify the cell clone with the earliest EMT. At the third passage, EMT began to occur in some cell clones. These early EMT cell clones were considered as BCSCs. EMT occurred randomly in wells, in which some cells changed shape from mesenchymal (Fig. 2A) to epithelial (Fig 2. B). This process continued until all cells showed epithelial-like shapes (Fig. 2C). We randomly selected one cell clone to analyze the characteristics of BCSCs. These procedures were performed similarly for all three samples.

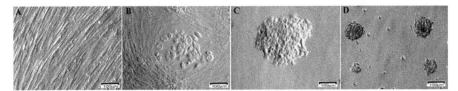

Figure 2. EMT of primary cells. Nearly 100% of cells exhibited a mesenchymal-like shape (A) when cultured in DMEM/F12 supplemented with 10% FBS. EMT of some cells (B) formed an islet of epithelial-like cells. Transitioned cells rapidly proliferated as condensed colonies (C).

3.3. Immunophenotype, rhodamine 123 efflux and mammosphere formation

We randomly selected three cell clones from the three samples to analyze the immunophenotype, rhodamine 123 efflux and mammosphere formation. The results are shown in Fig. 3. BCSC candidates showed a highly homogenous CD44⁺CD24⁻ phenotype (Fig. 3A) with more than 98% positive (98.82±0.72%) (n=3). These cell clones also contained a subpopulation (SP- Rhodamine 123 efflux phenotype) that showed more than 66.64±8.51% (n=3) (Fig. 3B). In addition, cell clones could form mammospheres when cultured in serum-free medium (Fig. 3C). Compared with BCSCs from CD44⁺CD24⁻/ᵈⁱᵐ cell sorting, the cell population was only 90.10±4.12% CD24⁻ and 8.19±3.38% CD24ᵈⁱᵐ (Fig. 3D). Notably, CD44⁺CD24⁻/ᵈⁱᵐ-sorted BCSCs contained a smaller SP than single-cell sorted BCSCs (34.56±3.48% vs. 66.64±8.51%) (Fig. 3E). However, the two kinds of BCSCs strongly exhibited mammosphere formation (Fig. 3F).

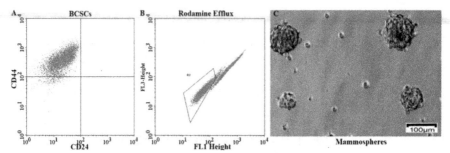

Figure 3. Flow cytometric analyses of immunophenotype and rhodamine 123 efflux. Cell clones exhibited the characteristics of BCSCs with near 100% CD44+CD24- (A), and were more than 50% SP-positive (D). Marker expression and the SP decreased in CD44+CD24-/dim-sorted cells. (D; E). However, both types of sorted cells could form mammospheres in serum-free medium (C, F).

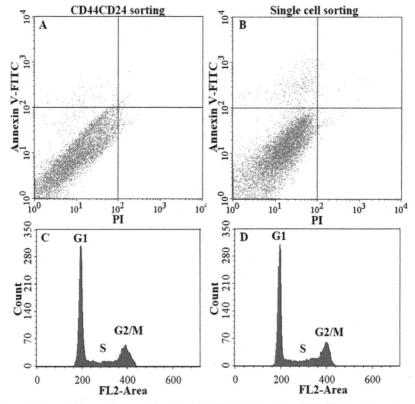

Figure 4. Doxorubicin resistance and the cell cycle of BCSCs isolated by CD44+CD24--sorting and single-cell sorting. At 3 μg/ml doxorubicin, BCSCs obtained from single-cell sorting did not undergo apoptosis (A), whereas apoptosis was observed among CD44+CD24- BCSCs (B). However, the cell cycles of these two populations were not significantly different (C–D).

3.4. Doxorubicin resistance and the cell cycle

BCSCs sorted by two strategies were evaluated in this study. BCSCs from single-cell sorting could resist doxorubicin more than CD44$^+$CD24$^{-/dim}$-sorted BCSCs. At 0 and 1 µg/ml doxorubicin, cells did not undergo apoptosis. However, at 3 µg/ml doxorubicin, no BCSCs underwent apoptosis from single-cell sorting, but there were 2.54±1.29% apoptotic CD44$^+$CD24$^{-/dim}$ BCSCs (Fig. 4A, B). Although these results showed that there was no significant different between BCSCs obtained from single-cell sorting and CD44$^+$CD24$^{-/dim}$-sorting (Fig. 4C, D).

3.5. *In vivo* tumor formation

Tumorigenicity is an essential characteristic of cancer stem cells. In almost all studies, tumor formation at a low number of injected cells is considered as the gold standard for cancer stem cell confirmation. In this study, we injected $1×10^3$, 10^4 and 10^5 cells into the mammary pad of NOD/SCID mice. At $1×10^3$ cells per mouse, BCSCs were able to form tumors (Fig. 5A, B). To

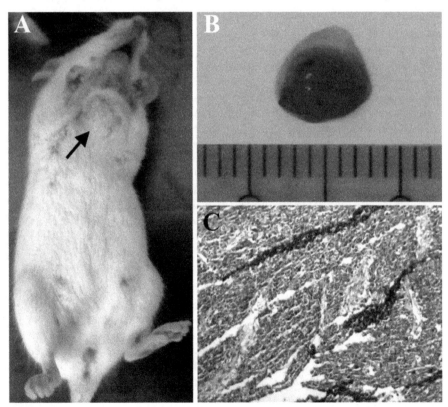

Figure 5. Tumor formation in NOD/SCID mice after injection of $1×10^3$ BCSCs. Tumors formed subcutaneously (A) at sizes from 5×6 mm (B), which were analyzed histochemically by HE staining (C).

confirm the histopathology of tumors, 10 μm tumor sections were stained with hematoxylin-eosin (HE). As shown in Figure 5C, tumors exhibited cancer cells with large nuclei. This result was similar to that of BCSCs from CD44+CD24-/dim-sorting in our previous studies [41]; [42]; [43].

4. Discussion

BCSCs are the origin of breast tumors. Thus far, the aim of many studies has been BCSC targeting. Recently, preclinical trials have demonstrated that agents targeting BCSCs are more effective than those targeting tumor cells. In all BCSC-based therapies, BCSCs are important for tumor targeting. However, the procedures used to isolate BCSCs are complicated and time consuming. Moreover, all present protocols obtain heterogenous populations of BCSCs. Indeed, sorting CD44+CD24- cells, a population considered as BCSCs, can express various levels of CD44 and CD24. SPs are also considered as BCSCs, but only a subpopulation show a CD44+CD24- phenotype. Thus, the aim of this study was to establish a new protocol to isolate homogenous BCSCs.

Similar to other techniques, our technique also cultured primary cells from malignant tumors as a first step. We successfully cultured 3/5 samples under this condition. Cells rapidly expanded around tumor fragments after 2 weeks. Various cell types appeared in primary culture, including mainly epithelial-like and mesenchymal-like cells. These results were consistent with our previous studies [41]; [42]. The primary cell population may contain at least six cell types including fibroblasts, mammary epithelial cells, mammary epithelial stem cells, breast cancer stem cells, breast cancer cells and stromal cells. To isolate homogenous BCSCs at a high purity, we applied single-cell sorting to individually isolate single cells in the wells of 96-well plates. For each sample, we sorted single cells into 2880 wells in 30 plates. From such a plate, we derived 14.67±5.13 cell clones, while other cells could not proliferate or died. Using this method, we can select cells that rapidly proliferate and survive when cultured alone. Indeed, normal cells and differentiated cells hardly proliferate when seeded as single cells. We considered that BCSCs existed among these clones. By changing the medium from M171 medium containing MEGS to DMEM/F12 supplemented with 10% FBS to select mesenchymal-like cell clones, all cell clones with an epithelial phenotype died and cell clones with a mesenchymal phenotype survived and proliferated.

There were 6.33±3.06 surviving cell clones per plate, with a mesenchymal-like shape. They were continuously subcultured for three to five passages, and cell clones that underwent the earliest EMT were chosen for further study. The results showed that cell clones will the earliest EMT occurred at the third passage. EMT resulted in cells spreading out to form an area with epithelial-like cells among mesenchymal-like cells. After 72 h, all cells transitioned into epithelial cells. We randomly selected three cell clones from a sample to analyze the characteristics of BCSCs. All three cell clones after EMT showed the properties of BCSCs. The cell population exhibited the common CD44+CD24- phenotype of BCSCs at 98.82±0.72%. This population was used to evaluate the SP by rhodamine 123 efflux. The results showed that the SP was CD44+CD24-/dim. To assess the multidrug-resistance property, we checked and compared with two other techniques, we recognized that BCSCs from single-cell sorting with higher antitumor drug resistance but the same tumor formation in NOD/SCID mice compared with CD44+CD24-/dim-sorting.

Single-cell sorting combined with subculture to isolate EMT cell clones exhibited several benefits for selection of BCSCs. Indeed, single-cell cloning is considered as the best technique for selection of a homogenous cell population. This technique is popular for cell cloning of hybridomas for monoclonal antibody production. However, in monoclonal antibody production, almost all studies use limited dilution or ring/syringe isolation. These two techniques have some limitations; particularly the low efficiency in dilution to obtain single cells in each well and it is laborious, time-consuming and uneconomical to screen samples with a low concentration of desired cells. Using the index sorting of the FACSJazz, it is easy to seed one cell in each well of 96-well plates. Single-cell sorting offers a new tool to efficiently and rapidly perform cell cloning.

In the next step, we cultured single cells to obtain cell clones for subculture. In mammalian cell culture, single-cell culture is usually suitable for transformed cells and immortal cell lines. Normal cells usually undergo apoptosis after 50±10 divisions because of the Hayflick limitation [27]. Thus, single-cell culture is only suitable for immortal cells. Indeed, in our study, when primary cells from tumors were individually cultured, some types of non-immortal cells can be eliminated after culture and subculture. Using single cell-based culture and sub-culture, almost all cell clones of stromal, epithelial and breast cancer cells can be lost over time. Thus, after three passages, some cell clones survive and can form continuous cultures. There were two cell types, namely epithelial-like cells and mesenchymal-like cells. We considered that mammary epithelial stem/progenitor cells exhibited the epithelial shape and breast cancer stem cells exhibited the mesenchymal shape. Mammary epithelial stem cells cannot survive in medium without hydrocortisone and EGF, whereas BCSCs do not depend on hydrocortisone or EGF [18]. Moreover, serum can inhibit the growth of normal epithelial cells from mammary tumors [17]. To eliminate mammary epithelial stem/progenitor cells, we changed the culture medium to DMEM/F12 supplemented with 10% FBS. After 48 h culture in this new medium, all epithelial-like cell clones did not survive, while all mesenchymal-like cell clones survived.

We propose that at this step, we successfully selected BCSC clones or BCSC-like clones. In the next experiment, we selected the strongest BCSC clone. The strongest BCSC clone was selected based on the time of EMT. EMT is related to the initiation of metastasis and cancer by BCSCs [2]; [52]; [20]; [53]; [48]; [45]; [35]. EMT can result in cells with stem cell properties [33]. Notably, Morel et al. (2008) could obtain BCSCs from EMT [37]. In a recent study, Blick et al. (2010) determined that EMT occurs together with the CD44+CD24-/dim phenotype of BCSCs [6]. Thus, in this study, cell clones that rapidly transitioned from mesenchymal to epithelial phenotypes were chosen as the most appropriate BCSC clone.

However, based on immunophenotypic and rhodamine 123 analyses, we found that BCSCs from single-cell sorting might not be homogenous. Indeed, a few cells did not exhibit the

CD44+CD24- immunophenotype or SP phenotype, indicating that BCSCs had differentiated into other cell types during proliferation. Such differentiation may be induced by medium containing serum. Some studies show that serum can induction differentiation of BCSCs [47]; [23]. Moreover, single cell-based culture easily induces differentiation [36]. Thus, single-cell sorting can obtain a pure population of BCSCs that must be maintained in a suitable medium to inhibit spontaneous differentiation.

5. Conclusion

Single-cell sorting is suitable for isolation of BCSCs to obtain a homogenous population for further experimentation and BCSC-targeting therapies. BCSCs obtained by this technique exhibit high purity, high resistance to doxorubicin, and form tumors in NOD/SCID mice at a low cell number. Compared with CD44+CD24- sorting, mammosphere culture and SP-based sorting, single-cell sorting in combination with subculture enables selection of EMT cell clones that give rise to a BCSC population with advantages such a homogenous population, higher doxorubicin resistance and mammosphere formation at high levels. However, spontaneous differentiation in culture is a problem that needs to be addressed. Single-cell sorting offers a new technique to detect and isolate BCSCs as well as other cancer stem cell types.

Author details

Phuc Van Pham, Binh Thanh Vu, Nhan Lu Chinh Phan,
Thuy Thanh Duong, Tue Gia Vuong and Ngoc Kim Phan
Laboratory of Stem Cell Research and Application, University of Science, Vietnam National University, Ho Chi Minh City, Vietnam

Giang Do Thuy Nguyen, Thiep Van Tran, Dung Xuan Pham and Minh Hoang Le
Ho Chi Minh City Oncology Hospital, Ho Chi Minh City, Vietnam

Acknowledgement

This work was funded by grants from the Vietnam National Project about Breast Cancer, Ministry of Science and Technology, Vietnam (Code: DTDL.2011-T/30). We thank the Oncology Hospital at Ho Chi Minh for supplying the malignant breast cancer tumors.

6. References

[1] Abraham BK, Fritz P, McClellan M, Hauptvogel P, Athelogou M, Brauch H. Prevalence of CD44+/CD24-/low cells in breast cancer may not be associated with clinical outcome but may favor distant metastasis. Clin Cancer Res 2005; 11: 1154-1159.

[2] Al-Hajj M, Wicha MS, Benito-Hernandez A, Morrison SJ, Clarke MF. Prospective identification of tumorigenic breast cancer cells. Proc Natl Acad Sci USA. 2003; 100(7): 3983-3988.

[3] Antón Aparicio LM, Cassinello Espinosa J, García Campelo R, Gómez Veiga F, Díaz Prado S, Aparicio Gallego G. Prostate carcinoma and stem cells. Clin Transl Oncol. 2007 Feb; 9(2):66-76.

[4] Avigan D, Vasir B, Gong J, Borges V, Wu Z, Uhl L, Atkins M, Mier J, McDermott D, Smith T, Giallambardo N, Stone C, Schadt K, Dolgoff J, Tetreault JC, Villarroel M, Kufe D. Fusion cell vaccination of patients with metastatic breast and renal cancer induces immunological and clinical responses. Clin Cancer Res. 2004; 10(14): 4699-708.

[5] Avigan D. Fusions of breast cancer and dendritic cells as a novel cancer vaccine. Clin Breast Cancer. 2003; 3(S4): 158-163.

[6] Blick T, Hugo H, Widodo E, Waltham M, Pinto C, Mani SA, Weinberg RA, Neve RM, Lenburg ME, Thompson EW. Epithelial mesenchymal transition traits in human breast cancer cell lines parallel the CD44^{hi}/CD24$^{lo/-}$ stem cell phenotype in human breast cancer. *J Mammary Gland Biol Neoplasia* 2010; 15: 235-252.

[7] Ceder JA, Jansson L, Ehrnström RA, Rönnstrand L, Abrahamsson PA. The characterization of epithelial and stromal subsets of candidate stem/progenitor cells in the human adult prostate. Eur Urol. 2008; 53(3): 524-31.

[8] Chen J, Chen ZL. Technology update for the sorting and identification of breast cancer stem cells. Chin J Cancer. 2010; 29(3): 265-9.

[9] Cioce M, Gherardi S, Viglietto G, Strano S, Blandino G, Muti P, Ciliberto G. Mammosphere-forming cells from breast cancer cell lines as a tool for the identification of CSC-like- and early progenitor-targeting drugs. Cell Cycle. 2010; 9(14): 2878-87.

[10] Colleoni M, Viale G, Zahrieh D, Pruneri G, Gentilini O, Veronesi P, Gelber RD, Curigliano G, Torrisi R, Luini A, Intra M, Galimberti V, Renne G, Nolè F, Peruzzotti G, Goldhirsch A. Chemotherapy is more effective in patients with breast cancer not expressing steroid hormone receptors: a study of preoperative treatment. Clin Cancer Res. 2004; 10(19): 6622-6628.

[11] Dontu G, Al-Hajj M, Abdallah WM, Clarke MF, Wicha MS (2003). Stem cells in normal breast development and breast cancer. Cell Prolif. 2003; 36(S1): 59-72.

[12] Dontu G, Jackson KW, McNicholas E, Kawamura MJ, Abdallah WM, Wicha MS. Role of Notch signaling in cell-fate determination of human mammary stem/progenitor cells. Breast Cancer Res. 2004; 6(6): R605-15.

[13] Engelmann K, Shen H, Finn OJ. MCF7 side population cells with characteristics of cancer stem/progenitor cells express the tumor antigen MUC1. Cancer Res. 2008; 68(7): 2419-2426.

[14] Eramo A, Lotti F, Sette G, Pilozzi E, Biffoni M, Di Virgilio A, Conticello C, Ruco L, Peschle C, De Maria R. Identification and expansion of the tumorigenic lung cancer stem cell population. Cell Death Differ 2008; 15: 504-514.

[15] Ferrandina G, Legge F, Mey V, Nannizzi S, Ricciardi S, Petrillo M, Corrado G, Scambia G. A case of drug resistant clear cell ovarian cancer showing responsiveness to gemcitabine at first administration and at re-challenge. Cancer Chemother. Pharmacol. 2007; 60(3): 459-61.

[16] Fillmore CM, Kuperwasser C. Human breast cancer cell lines contain stem-like cells that self-renew, give rise to phenotypically diverse progeny and survive chemotherapy. Breast Cancer Res. 2008; 10(2): R25.

[17] Gaffney EV, Pigott D. Effect of serum on cells cultured from human mammary tumors. In Vitro. 1978; 14(5): 451-457.

[18] Gaffney EV, Pigott D. Hydrocortisone stimulation of human mammary epithelial cells. In Vitro. 1978; 14(7): 621-624.

[19] Gao SM, Yang J, Chen C, Zhang S, Xing CY, Li H, Wu J, Jiang L. miR-15a/16-1 enhances retinoic acidmediated differentiation of leukemic cells and is upregulated by retinoic acid. Leuk Lymphoma. 2011; 52(12): 2365-2371.

[20] Ginestier C, Hur MH, Charafe-Jauffret E, Monville F, Dutcher J, Brown M, Jacquemier J, Viens P, Kleer CG, Liu S, Schott A, Hayes D, Birnbaum D, Wicha MS, Dontu G. ALDH1 is a marker of normal and malignant human mammary stem cells and a predictor of poor clinical outcome. *Cell Stem Cell* 2007; 1(5): 555-67.

[21] Glinsky GV. Stem cell origin of death-from-cancer phenotypes of human prostate and breast cancers. Stem Cell Rev. 2007; 3(1): 79-93.

[22] Gonzalez-Angulo AM, Morales-Vasquez F, Hortobagyi GN. Overview of resistance to systemic therapy in patients with breast cancer. Adv Exp Med Biol. 2007; 608: 1-22.

[23] Grimshaw MJ, Cooper L, Papazisis K, Coleman JA, Bohnenkamp HR, Chiapero-Stanke L, Taylor-Papadimitriou J, Burchell JM. Mammosphere culture of metastatic breast cancer cells enriches for tumorigenic breast cancer cells. Breast Cancer Res. 2008; 10(3): R52.

[24] Guarneri V, Conte PF. The curability of breast cancer and the treatment of advanced disease. Eur J Nucl Med Mol Imaging. 2004; 31(S1): S149-161.

[25] Guo J, Zhou J, Ying X, Men Y, Li RJ, Zhang Y, Du J, Tian W, Yao HJ, Wang XX, Ju RJ, Lu WL. Effects of stealth liposomal daunorubicin plus tamoxifen on the breast cancer and cancer stem cells. J Pharm Pharm Sci. 2010; 13(2): 136-151.

[26] Hata Y, Etoh T, Inomata M, Shiraishi N, Nishizono A, Kitano S. Efficacy of oncolytic reovirus against human breast cancer cells. Oncol Rep. 2008; 19(6): 1395-1398.

[27] Hayflick L. The cell biology of aging. J Invest Dermatol. 1979; 73(1): 8-14.

[28] Hiraga T, Ito S, Nakamura H. Side population in MDA-MB-231 human breast cancer cells exhibits cancer stem cell-like properties without higher bone-metastatic potential. Oncol Rep. 2011; 25(1): 289-96.

[29] Kochhar RKV, Bejjanki H, Caldito G. Statins reduce breast cancer risk: A case control study in US sfemale veterans. J Clin Oncol. 2005; 23: 6s.

[30] Korkaya H, Paulson A, Charafe-Jauffret E. Regulation of mammary stem/progenitor cells by PTEN/Akt/beta-catenin signaling. PLoS Biol. 2009; 7: e1000121.

[31] Korkaya H, Paulson A, Iovino F. HER2 regulates the mammary stem/progenitor cell population driving tumorigenesis and invasion. Oncogene 2008; 27: 6120-6130.

[32] Li C, Heidt DG, Dalerba P, Burant CF, Zhang L, Adsay V, Wicha M, Clarke MF, Simeone DM. Identification of pancreatic cancer stem cells. Cancer Res. 2007; 67: 1030-1037.

[33] Mani SA, Guo W, Liao MJ, Eaton EN, Ayyanan A, Zhou AY, Brooks M, Reinhard F, Zhang CC, Shipitsin M, Campbell LL, Polyak K, Brisken C, Yang J, Weinberg RA. The epithelial-mesenchymal transition generates cells with properties of stem cells. *Cell* 2008; 133: 704-715.

[34] Marcato P, Dean CA, Giacomantonio CA, Lee PW. Oncolytic reovirus effectively targets breast cancer stem cells. Mol Ther. 2009; 17(6): 972-9.

[35] McDermott SP, Wicha MS. Targeting breast cancer stem cells. *Mol Oncol* 2010; 4: 404-419.

[36] McFarland DC. Preparation of pure cell cultures by cloning. Methods Cell Sci. 2000; 22(1): 63-66.

[37] Morel AP, Lievre M, Thomas C, Hinkal G, Ansieau S, Puisieux A: Generation of breast cancer stem cells through epithelial-mesenchymal transition. *PLoS One* 2008; 3: e2888.

[38] Parkin DM, Bray F, Ferlay J, Pisani P. Global cancer statistics. CA Cancer J Clin. 2005; 55(2): 74-108.

[39] Parkin DM, Fernández LM. Use of statistics to assess the global burden of breast cancer. Breast J. 2006; 12(S1)1: S70-80.

[40] Pham Van Phuc, Chi Jee Hou, Nguyen Thi Minh Nguyet, Duong Thanh Thuy, Le Van Dong, Truong Dinh Kiet and Phan Kim Ngoc. Effects of breast cancer stem cell extract primed dendritic cell transplantation on breast cancer tumor murine models. Annual Review & Research in Biology 2001; 1(1): 1-13.

[41] Pham Van Phuc, Tran Thi Thanh Khuong, Le Van Dong, Truong Dinh Kiet, Tran Tung Giang and Phan Kim Ngoc. Isolation and characterization of breast cancer stem cells from malignant tumours in Vietnamese women. JCAB 2010; 4(12): 163–166.

[42] Phuc PV, Nhan PLC, Nhung NT, Nhung TH, Thuy DT, Dong LV, Kiet TD and Ngoc PK. Differentiation of breast cancer stem cells by knockdown of CD44: promising differentiation therapy. Journal of Translational Medicine 2011; 9(1): 209.

[43] Phuc PV, Nhan PLC, Nhung TH, Tam NT, Hoang NM, Tue VG, Thuy DT, Ngoc PK. Downregulation of CD44 reduces doxorubicin resistance of CD44+CD24- breast cancer cells. OncoTargets and Therapy 2011; 4: 71-7.

[44] Pollock CB, Koltai H, Kapulnik Y, Prandi C, Yarden RI. Strigolactones: a novel class of phytohormones that inhibit the growth and survival of breast cancer cells and breast cancer stem-like enriched mammosphere cells. Breast Cancer Res Treat. 2012 Mar 29. [Epub ahead of print]

[45] Polyak K, Weinberg RA: Transitions between epithelial and mesenchymal states: acquisition of malignant and stem cell traits. *Nat Rev Cancer* 2009; 9: 265-273.

[46] Pommier SJ, Quan GG, Christante D, Muller P, Newell AE, Olson SB, Diggs B, Muldoon L, Neuwelt E, Pommier RF. Characterizing the HER2/neu status and metastatic potential of breast cancer stem/progenitor cells. Ann Surg Oncol. 2010; 17(2): 613-23.

[47] Ponti D, Costa A, Zaffaroni N, Pratesi G, Petrangolini G, Coradini D, Pilotti S, Pierotti MA, Daidone MG. Isolation and in vitro propagation of tumorigenic breast cancer cells with stem/progenitor cell properties. Cancer Res. 2005; 65(13): 5506-5511.

[48] Prat A, Parker JS, Karginova O, Fan C, Livasy C, Herschkowitz JI, He X, Perou CM: Phenotypic and molecular characterization of the claudin-low intrinsic subtype of breast cancer. *Breast Cancer Res* 2010; 12: R68.

[49] Prince ME, Sivanandan R, Kaczorowski A, Wolf GT, Kaplan MJ, Dalerba P, Weissman IL, Clarke MF, Ailles LE. Identification of a subpopulation of cells with cancer stem cell properties in head and neck squamous cell carcinoma. Proc. Natl. Acad. Sci. USA 2007; 104: 973-978.

[50] Sansone P, Storci G, Giovannini C. P66Shc/Notch-3 interplay controls self-renewal and hypoxia survival in human stem/progenitor cells of the mammary gland expanded in vitro as mammospheres. Stem Cells 2007; 25: 807-815.

[51] Seo DC, Sung JM, Cho HJ, Yi H, Seo KH, Choi IS, Kim DK, Kim JS, El-Aty AM A, Shin HC. Gene expression profiling of cancer stem cell in human lung adenocarcinoma A549 cells. Mol. Cancer 2007; 6: 75.

[52] Sheridan C, Kishimoto H, Fuchs RK, Mehrotra S, Bhat-Nakshatri P, Turner CH, Goulet R Jr, Badve S, Nakshatri H. CD44+/CD24- breast cancer cells exhibit enhanced invasive properties: an early step necessary for metastasis. Breast Cancer Res. 2006; 8(5): R59.

[53] Thiery JP, Acloque H, Huang RY, Nieto MA: Epithelial-mesenchymal transitions in development and disease. *Cell* 2009; 139: 871-890.

[54] Xie G, Yao Q, Liu Y, Du S, Liu A, Guo Z, Sun A, Ruan J, Chen L, Ye C, Yuan Y. IL-6-induced epithelial-mesenchymal transition promotes the generation of breast cancer stem-like cells analogous to mammosphere cultures. Int J Oncol. 2012; 40(4): 1171-1179.

Specific Uses of Tissue Culture

Chapter 6

Use of Cell Culture to Prove Syncytial Connection and Fusion of Neurons

O.S. Sotnikov

Additional information is available at the end of the chapter

1. Introduction

1.1. Discussion between neuronists and reticularists

We do not consider it possible to cast any doubts on the grounds of the Neuronal Doctrine; however, the interneuronal syncytial connection still does exist and the cytoplasmic fusion of neurons still is possible.

The reticular theory of the general syncytial cytoplasmic connection of neuronal processes as the principle of organization of the nervous system was already known to the German histologist Joseph von Gerlach [1]. It was supported by almost all neurologists of the XIX century [2] and was passionately defended by the famous Camillo Golgi.

The presence of the cytoplasmic syncytium[1] in the nervous system was defended by such known histologists as Nissl, Ranvier, Schwann, and others. The theory of nervous network had its attractive and convenient explanations [2]. By suggesting the general cytoplasmic connection of nerve fibers, it considered nervous network as anastomoses, roundabout pathways of blood vessels and allowed explaining comparatively easily the relatively fast recovery of functions in cerebral stroke. The reticular theory suggested not the discrete single, but the grouped functioning of neurons, which at present seems more realistic [3, 4]. However, this theory was not based on the most important — absolute scientific facts. The wonderful histological method invented by C. Golgi (*reazione nera*) and celebrated by S. Ramon y Cajal could not compensate low resolving power of the optic microscope. Not infrequently the superpositions of nerve processes one upon another were interpreted as their fusion and formation of network. The reticular theory had to be replaced by the Neuronal Doctrine.

---

[1] Here and further the term "syncytium" implies the true cytoplasmic connection of different neurons or different processes of one neuron rather than the quasisyncytium representing the interneuronal electric connection with aid of gap junction. The term "syncytium" is incorrectly borrowed by many electrophysiologists [15].

Lucky findings of mossy and climbing fibers by S. Ramon y Cajal proved convincingly the existence of nerve terminals and a possibility of individual existence of neurons. Then S. Ramon y Cajal and his supporters obtained numerous preparations in different parts of the nervous system, which convinced scientific community in the rightness of the Neuronal Doctrine of W. Waldeyer and S. Ramon y Cajal [5, 6]. Disputes of Cajal and Golgi represented one of the largest collisions of ideas in development of the scientific thought [7].

Discussion of neurohistologists was quite emotional. Santiago Ramon y Cajal thought that – hypothesis of the network is a terrible enemy, – contagion of reticularism, whereas Camillo Cajal opposed to him defiantly emotionally his views in his Nobel lecture [2].

It seemed that discovery of synapses with aid of electron microscope [8, 9] had become the absolute and last proof of victory of the Neuronal Doctrine. In most neurobiologists the erroneous opinion was formed about the absence in principle of the syncytial connection in the nervous system. However, the proof of the presence of synapses, strictly speaking, is not the proof of the absence of the syncytial interneuronal connection. This is the typical error (paralogism), a usual contrivance in discussions of sophists of the IV century B.C. [10].

All neurons have synapses (incomplete premise).
Synapse is the form of connection of neurons.
Hence, the form of connection of all neurons is synapse.

Both reticularists and neuronists assumed the exclusively one way of connection in the nervous system. Both the former and the latter in principle did not tolerate compatibility of the theories. This was characteristic both of S. Ramon y Cajal and of C. Golgi (Neuronismo o Reticularismo), either the neuronal or the reticular theory [11]. However, this classical approach in discussions "either – or" already at that time could have been replaced by the approach "both – and", as the third opinion already existed [12]. Some researchers adhering to the Neuronal Theory noted a possibility of the existence of the interneuronal syncytium in particular places and under some conditions [13, 14]. Even C. Golgi extended the theory of network only to organization of axonal branchings, whereas dendrites, in his opinion, terminated freely [2]. At present, in the literature there is the sufficient amount of evidences for accepting the concept and proofs of reticularists erroneous and nevertheless for claiming that the syncytial connection between neurons does exist.

2. Findings of syncytial connection in the nervous system

However, in the literature there already are the irrefutable facts of the presence in the nervous system of the true cytoplasmic syncytial interneuronal connection. After information of absolute facts of fusion of nerve processes in invertebrates [16], some authors were ready to recognize the giant neurons to be the non-nerve cells rather than to agree with facts of their syncytial connection [17]. It is impossible to ignore detection of the syncytial connection in molluscs, crustaceans, polychaetes, and other invertebrates [18-24]. All these works first of all offer absolute proofs of that the interneuronal syncytium in the nervous system in principle does exist.

Data about "fused neurons, that produce the giant axons...", are "example of a situation, that is against the strict letter of the neuron doctrine but can fit easily into the cell theory" [25]. It turns out that in nature the presence of chemical synapses in animals is quite compatible at solution of special tasks with the presence of the cytoplasmic syncytial connection. In nature the Neuronal Doctrine and structural elements of the reticular theory are compatible.

Of principal importance are the data that syncytial connections can also be formed between stumps of sectioned fiber at its regeneration by ingrowth of the central stump into the peripheral one [26-28]. By the example of formation of the earthworm giant axon, in some cases, the "Calenary theory" was even demonstrated to be true [29]; this theory was a variant of the syncytial fusion of neurons and suggested formation of nerve fibers by fusion of individual nerve cells into chains. This theory was held by one of pillars of neuronism van Gehuhten [2].

We were the first to reveal with aid of electron microscope the syncytial connection *in situ* in the piglet enteral nervous system [30]. Subsquently, this way of interneuronal connection was shown in molluscs [31, 32], in the crawfish peripheral nervous system [33], in cat autonomic ganglia [34], in rabbit hippocampus [35, 36], and in the human cerebral cortex [12, 37].

Morphology of fusion of living neurons at present has not yet been studied, and to do this, it is the most convenient with use of tissue culture.

3. Obtaining and cultivation of isolated mollusc neurons

By having obtained convincing data about the existence of interneuronal syncytial connection in invertebrates, we found it quite important to reveal such form of connection in living mollusc neurons in cell culture [38]. For this, the peripharyngeal ring of the mollusc *Lymnaea stagnalis* with all ganglia (Fig. 1) were performed in 0.4% pronase with temperature of 20°C (use of another protease is also possible). We used lyophilized pronase from *Streptomyces griseus* (Sigma). The mean molecular mass is 20000. The pronase solution was prepared on the basis of isotonic solution for molluscs.

Use of pronase for enzymatic treatment of ganglia was due to the following causes. Pronase represents a complex preparation composed of 11-14 proteases. Its hydrolytic activity is spread onto the large spectrum of proteins, peptides, amides, and amino acid esters [39]. Proteases are known to have properties of a neurotrophic factor. They stimulate growth and branching of neurites [40, 41]. M.A. Kostenko was the first to develop the method of proteolytic isolation of individual neurons with aid of pronase [42]. Subsequently this method of enzymatic dissociation has become widely used in electrophysiological and biophysical studies [43-46]. Other methods of cell disintegration also are known [47].

At treatment of ganglia with pronase, the earlier and greater damages are characteristic of elements of connective tissue and glia. This is due to that they are in the composition of sheaths and are the first to deal with acting factors. This also depends on that glial cells have very large surface of their lamellas [48] and accordingly the extremely high ratio of the membrane area to the cell volume (i.e., have very low thermodynamic and structural stability). They always are the first to respond to external actions. If time of the proteolytic treatment is decreased, some number of gliocytes can be preserved in culture.

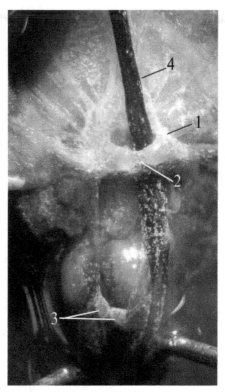

Figure 1. The peripharyngeal nerve ring of the mollusc *Lymnaea stagnalis* L *in vivo*.
1 – cerebral ganglion; 2 – intercerebral commissure; 3 – buccal ganglia; 4 – esophagus. MBS-2. Magnification – 20×.

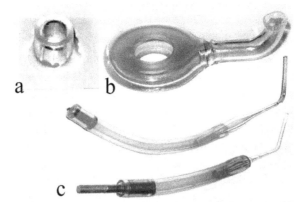

Figure 2. Microcameras (a, b) and Γ-shaped micropipettes (c), with whose aid isolated neurons were cultured and prepared.

After 40-60 min of incubation of ganglia in pronase, they were washed out from proteolytic enzymes in isotonic solution and got free from their connective tissue. To make sure in the viability of neurons, they were placed under a MSSO-IV42 microscope (LOMO, St. Petersburg) and by using a gold microelectrode with the tip diameter of about 20 μm in the glass isolation the extracellular leading of the spontaneous electric activity of individual cells was performed. They generated spikes with frequency of 1.5 ± 0.3 imp/s. The amplitude of spikes was equal to 22 ± 4 μV, while their duration – to 4 ms [38].

After cleaning from capsules, the procedure of multiple suction of ganglia (pipetting) into the specially designed Γ-shaped glass micropipettes with the tip diameter of 0.8 and 0.6 mm was performed (Fig. 2).

The experience shows that the molluscan neurons can be seeded onto the glass without special support. For cultivation of neurons we used microcameras of the following construction (Fig. 2, a, b).

1. The glass rings, 0.8 cm in height and 0.8 cm in diameter, attached with a mixture of paraffin and vaseline (1:1) to coverslips served the base of the camera.
2. The Carelli flask in our modification represents camera, 4 ml in volume, with two round holes on the upper and lower walls (Fig. 2, b), which for preservation of sterility were closed by coverslips and soldered with the paraffin—vaseline mixture. The cylindric camera branches designed for change of the nutrition medium were closed by a lid. The cells were seeded onto the lower coverslip, which allowed microscopy with aid of inverted microscope at high magnification (with objectives 40× and 60×). Besides, clearness and contrast of image were significantly improved. At filling of only the lower camera indentation (1.5-2 mm in depth) a meniscus of the medium was formed, and the surface tension produced aggregation of neurons near the camera center. Cells in this aggregate were not flattened, but could long survive by preserving the sphere-like shape and the granular cytoplasm. We managed to avoid aggregation of neurons at the complete filling of the camera with nutrition medium. The cells were attached to the surface of the lower coverslip, spread on it, and generated processes. Our proposed camera allows observing the details of growing neurons, invisible at low magnifications or resulted from light dispersion at its passage through the thick bottom of the glass or plastic Petri dish. Thus, for instance, it is clearly seen that the structures sometimes mistaken by researchers for cell processes are in reality dense cytoskeleton strands inside lamellas. For experimental studies, the serum-free nutrition medium of certain chemical composition was used. As the stock medium, we used the standard RPMI 1640 medium.

The thereby isolated cells were repeatedly washed with Ringer solution. The upfloating connective tissue and dying gliocytes were carefully removed with the same micropipette. As a result of micropipetting, we managed to obtain a significant amount of isolated glia-free neurons (Fig. 3). By varying pronase concentration, time of incubation, and activity of ganglion pipetting, it was possible to obtain either the isolated neuronal bodies or neurons with fragments of their processes. The more time of treatment with pronase and its concentration, the lower chances to isolate neuron with its processes.

The medium directly for cultivation was prepared based on the stock single medium RPMI 1640 medium by its dilution with a special salt solution. The composition of the salt solution (mol): 75 NaCl, 5 KCl, 2.5 CaCl$_2$, 2.5 MgCl$_2$. To obtain the nutritive medium, 250 ml of the single RPMI 1640 medium were diluted in 1 l of the salt solution. Concentration of amino acids after dilution decreased 5 times. Such medium composition is optimal for successful cultivation of dissociated neurons of *Lymnaea stagnalis*.

pH of the prepared medium raised to 7.6 with aid of Tris-HCl, and for sterilization filtrated immediately the medium by passing it through Millipore membranes with diameter of pores of 0.22 µm by using a Peristaltic miniflon pump type 304 (Poland). In the process of filtration, pH of medium reached the value of 7.8 optimal for cultivation. For control of pH, there was used a pH-meter-millivoltmeter of the "pH 150" type ("Izmeritel", Gomel).

The light microscopy was performed by using a MBI-13 inverted phase-contrast microscope (LOMO, St. Petersburg) with a thermostated camera and water-heat filter.

Observations were performed only on living neurons. The neuron viability signs are the clear external cell contour (the intact membrane), the granular cytoplasm that has the clear light dispersion and fills completely the body contour; the nucleus contours are not seen. Considered damaged were the neurons, in which the granular cytoplasm was separated at small sites from the outer cell body contour with a layer of homogenous flooded cytoplasm. Not infrequently the nucleus contours are well expressed. The hopelessly traumatized cells usually had the aggregated cytoplasm with large flooded submembranous spaces and the Brownian movement in them of granules or the visible disturbance of the membrane integrity. The sign of insufficient treatment with pronase served preservation of fragments of the glial membrane connected with the cell body. The neuron was as if in the villous membrane with uneven contours, while its own "membrane" was not seen.

Observation on behavior of regenerating processes can begin at the very first day; however, at long illumination under microscope, especially in the absence of water filter, neurites stop growing and branching.

Thus, the exposed detailed procedure allows obtaining the viable single neurons that regenerate nerve processes (Fig. 3, *d*) and form interneuronal contacts and extensive nerve plexuses. It is important to emphasize that we are dealing here with culturing of the glia-deprived brain neurons of the adult animals and that the culture medium is strongly identified and does not contain uncertain growth factors. This makes it possible to take into account sufficiently exactly the effects of various outer agents on living neurons, their processes, and interneuronal contact.

To reveal the principal possibility of formation of syncytium in the nervous system, it is necessary, first, to reproduce on living cells the process of fusion of outer cellular neuronal membranes and, second, to prove combination of the neuroplasm of two different cells. The most convenient for this is use of living neurons in the tissue culture.

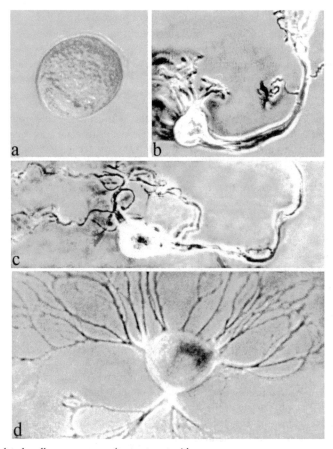

Figure 3. Isolated molluscan neurons after treatment with pronase.
a – neuron isolated without processes; b, c – variants of neurons isolated with processes; d – isolated neuron at the 4th day of cultivation in the RPMI 1640 medium. Supravital microscopy. Obj. 20Ph, eyep. 10.

4. Structural kinetics of neurons during formation of syncytium

The primary neuronal cultures were studied in phase contrast, which increased essentially the degree of detection of structural details with aid of a light microscope. However, the main virtue of the method of supravital studies was a possibility of studying the structural kinetics of the culture, the ability of the method to reveal structural transformation of neuron in time. This advantage of the method also allowed developing peculiar absolute criteria making it possible to prove formation of syncytial connections in living neurons at the light microscopy level [32, 49]. The studies were carried out by using the many-day (3-6 days) automatic time-lapse microvideo shooting and computer analysis. To prove formation of the neuronal cytoplasmic connection, we used criteria that allow with aid of videostudy of kinetics of living neurons differentiating fusion of processes from their contact. For this purpose, as the theoretical substantiation (criterion of syncytial connection) there was used inverted statement of the law of the Wallerian degeneration [50]. Astonishing as it is, detection of syncytium was helped by the concept of the Wallerian degeneration that at its time was one of important proofs of the absence of syncytial connection in the nervous system.

Since after separation of the nerve process from the neuronal body (the trophic center) it should necessarily degenerate, so if such process does not degenerate after separation of the body of its neuron, this means that it is in the cytoplasmic connection with body of the other neuron.

We were the first to propose the novel way of revealing syncytial connection between neurons by using the light microscopy observations of dynamics of structural processes in the tissue culture.

Figure 4 shows the initial establishment of "end-to-end" contacts between filopodia (3) of the lamellar processes of cells A and B. Then, after 15 min, contacts start to form between lamellae 1, 2 neighboring cells A, B and the boundary between them becomes unclear. It can be suggested that these lamellae have a syncytial coupling, as with time they form a single intracytoplasmic cytoskeletal filament (Fig. 4, d-f, 4) which, being continuous, runs from one lamella to other lamellae. Cell A is connected with cell B via lamella 2 of cell B (Fig. 4, d, 4). However, this remains insufficient as evidence of a syncytial coupling between these lamellar processes, though, as often occurs in primary cultures, cell B dies (Fig. 4, e, 5) and its lamellar processes 2 and cytoskeletal filament 4 remain intact. Lacking its body (the trophic center), process 2 of cell B does not undergo Wallerian degeneration, as occurs in all other cases when neuron processes are detached from their bodies. In the present case, they persist for 4 h, i.e., until the end of the observation period. Furthermore, by contracting, cells A and B are brought closer together (the distance between them decreases by 9.2% and the anastomosis connecting them straightens, Fig. 4, d, f). This is possible only in one case, if the process preserving viability got time to acquire a new trophic center, i.e., established the direct cytoplasmic connection with the other cell. This process is schematically presented in three stages (Fig. 5).

Exchange of neuroplasm can be observed after fusion of processes. Thus, the plexus of tubular nerve processes presented in Fig. 6, a, b shows that growing neurite 1 establishes contact with process 2 of the other cell. Further video frames show displacement of the cytoplasm of varicosity 3 of process 1 into process 2 of cell 4 (Fig. 6, b-f), which provides further support for the formation of syncytial connections between different cells by fusion of their processes. The cytoplasmic varicosity 3 of process 1 slowly, over 6 h 14 min, moves to the site of contact between processes 1 and 2 (Fig. 6, c) and, passing it (Fig. 6, d), moves from process 1 into process 2. The varicosity then approaches cell 4 (Fig. 6, e) and fuses with it (Fig. 6, f). It is suggested that displacement of the cytoplasmic varicosity from the process of one neuron to the process of the other one can only occur when there is a syncytial connection between the neuroplasm of their processes.

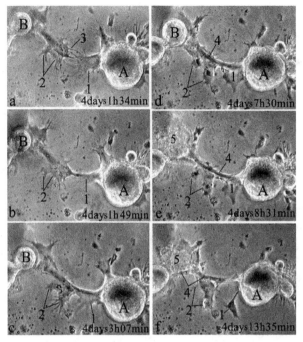

Figure 4. Dynamics of formation of syncytial connection between processes of neurons of the mollusc *Lymnaea stagnalis* in tissue culture.
a-f – stages of formation of syncytium (time – from beginning of culturing); A, B – neurons forming syncytium; 1 – process of neuron A; 2 – lamellar process of neuron B; 3 – combined filopodia of growth cones of cells A and B prior to fusion of their lamellar processes; 4 – the formed single cytoskeletal structure inserted into lamellopodia of both cells; 5 – dead neuron. Tissue culture, time-lapse computer videoshooting. Obj. 20Ph, eyep. 10.

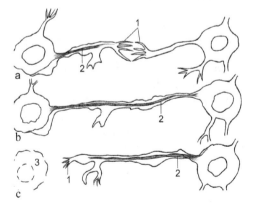

Figure 5. Schematic presentation of formation of syncytial collection between processes of two neurons.
a-c – stages of process; 1 – filopodia of growth cones; 2 – formation of the cytoskeletal strand common to two neurons; 3 – the dead neuron.

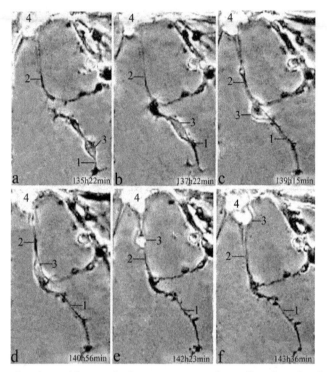

Figure 6. Formation of syncytial connection between processes of two cells and translocation of the cytoplasmic varicosity from one process into the other.
a-f – stages of formation of syncytium; 1 – process of the lower cell; 2 – process of the upper cell; 3 – the varicosity that flows over through the place of fusion of two processes; 4 – body of the upper cell. Tissue culture, time-lapse computer videoshooting. Time from the beginning of culturing is shown. Obj. 20Ph, eyep. 10.

Such behavior of cytoplasmic varicosities, in our opinion, can serve another criterion of formation of the cytoplasmic syncytial connection of neurons.

In the dissociated culture, in isolated neurons that have no contacts with other neurons there are formed multiple contacts between their processes (autapses) (Fig. 7, a). Some of these processes seem to form syncytial connection, as cytoplasmic varicosities from some processes are freely translocated onto neighbor processes. Processes 1 and 2 (Fig. 7, b) approach the process 3, while their cytoplasmic varicosities overcome sites of their connections (Fig. 7, d-f). This phenomenon occurs for 5 h between all processes of the isolates neuron

Such are the facts of fusion of neurons with formation of the syncytial interneuronal connection in tissue culture. However, by the example of cells of other types it is shown that their syncytial connection is easily transformed into the cell fusion. Dynamics of fusion of neuronal bodies has not yet been studied at present.

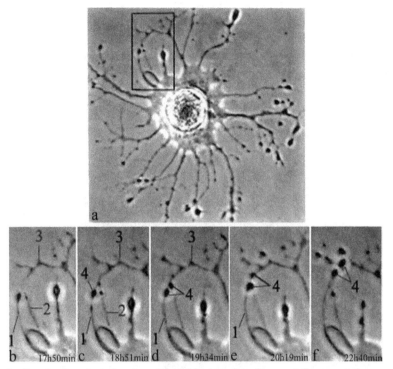

Figure 7. Translocation of cytoplasmic varicose thickenings along the syncytially connected processes of the same neuron (a).
b-f – stages of overflowing of cytoplasmic thickenings, 1, 2 – nerve processes of the lower neuron, 3 – process of the upper neuron, 4 – cytoplasmic varicosities translocating from the lower to the upper processes. Tissue culture, time-lapse computer videoshooting. Time – from the beginning of culturing. Obj. 20Ph, eyep. 10.

5. Spontaneous syncytial fusion of bodies of molluscan living neurons

It seemed unachievable to wish obtaining in the supravital experiment the fusion of living neuron bodies with formation of binucleated cells. However, numerous experiments on proteolytic dissociation of ganglia with obtaining isolated neurons got success several times. Preparations of living, just isolated mollusc cells turned out to contain binucleated neurons. The fact that indeed these were individual binucleated, rather than attached paired cells is indicated, first, by the treatment itself of cells with pronase that regularly, in 100% of cases, dissociates molluscan ganglia into single neurons by eliminating all gliocytes and fibroblasts. Second, these cells sometimes had common fused nerve processes (Fig. 8). Lastly, the angle between the fused cells (the angle of waist) exceeded 125°, which has been proven electrophysiologically [51] to be the convincing evidence for fusion of neurons.

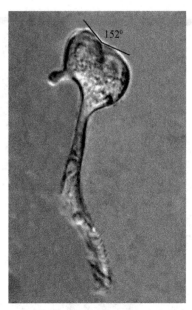

Figure 8. Spontaneous fusion of two neurons isolated from the molluscan perinuclear ring. Obj. 40Ph, eyep. 10.

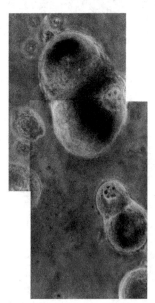

Figure 9. Binucleated neurons of *Lymnaea stagnalis* after two days of incubation in the cultural RPMI 1640 medium. Phase contrast. Obj. 20Ph, eyep. 10.

Unfortunately, the revealed fused neurons and the very process of the cell fusion directly after isolation are observed quite seldom. However, after careful washing out from pronase, after two days of incubation of neurons in the culturing medium there is noted a significant number of the fused neurons whose fusion angles are much higher than 125º (Fig. 9). Such cells also have other absolute features of fusion.

This has allowed us to suggest that neurons, by using their own potentials for fusion, some time after isolation, without help of draining agents, are able to form intersomatic syncytial cytoplasmic connections and to fuse. But since in the experiment there was involved a complex of proteolytic enzymes (0.4% pronase), a possibility existed that the cell fusion had been provoked by proteolytc enzymes. To check this, we performed a special series of experiments on electron microscopy study of effect of pronase on molluscan neurons.

Nowhere in this experiments could we detect formation of typical specialized membranous contacts and syncytial perforations. It can be concluded that the treatment itself of neurons with pronase does not induce their fusion. It only removes glial interlayers between the nerve cell bodies by promoting the fusion.

6. Development of method of artificial syncytial cytoplasmatic fusion of neurons *in vitro*

Our application of several procedures used at fusion of non-nervous cells [52, 53] had no success. Attempts at using polyethyleneglycol as a draining agent turned out to be unsuccessful, as these neurons do not endure the temperature necessary for keeping polyethyleneglycol in melted state. For fusion of neurons with aid of latex balls of polysterol there was needed the culture medium deprived of Ca and Mg ions causing aggregation of balls. Such medium is also poorly endured by molluscan neurons. As a result, these methods turned out to be poorly effective for the studied neurons.

At developing the new way [54], two groups of experiments were performed. In the first group, isolated neurons were studied in usual culture for 2-5 days, while in the second group, the structural processes were analyzed in the cell aggregate after two days.

Since earlier nobody has performed the experimental fusion of non-infected neurons, we developed a special method for fusion of the molluscan cells that have the satellite gliocytes.

Ganglion neurons of the mollusc *Limnaea stagnalis* first were freed from the ganglion connective tissue and satellite gliocytes with aid of proteolytic treatment. Neurons were carefully washed out from pronase. Then they were studied in the Eagle MEM cultural medium (Sigma, England). A part of cells were aggregated by centrifugation (3000 *g*, 15 min) and preserved in the cultural nutritive medium for 2 days by allowing the neurons to restore their natural capabilities for adhesion and fusion. Then, with aid of a light phase-contrast microscope, semithin sections were analyzed, while with aid of the standard transmissional electron microscope, ultrastructure of borders of contacting neurons was studied.

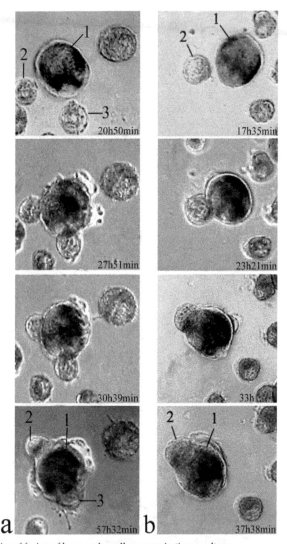

Figure 10. Dynamics of fusion of large and small neurons in tissue culture.
a – fusion of two neurons for 57 h; b – fusion of three neurons for 37 h; 1, 2, 3 – fusing neurons. Time-lapse videoshooting. Obj. 20Ph, eyep. 10.

Experimental aggregates of neurons were carefully washed out from pronase with cultural medium and were kept under the "normal" conditions natural for cultivated neurons. Therefore, we consider the processes revealed in the aggregates as natural, depending only on potentials of the cells themselves. It is suggested that treatment with pronase promoted only the removal of glial membranes.

Under conditions of the performed experiment the single living neurons often form paired or multicellular aggregates and fuse between each other. Fig. 10 presents dynamics of fusion of neurons in tissue culture. Initially, neurons are approached to each other by using the dotted contact. Later, the area of contact is enlarged and acquires shape of the large flatness.The angle of waist, of the node between the fusing cells increases. At fusion of small neurons this phenomenon is expressed worse due to large curvature of small spheres. Lastly, the smaller neuron is invaginated almost completely into the larger cell.

At the second day of culturing, the neuronal processes start growing and provide contacts of neurons; by contracting, the processes make the neuronal bodies adjacent to each other (Fig. 11). The contacting neuronal bodies form the 8-like structures that are separated by vacuole-like structures (Fig. 11, c, d). These vacuole-like structures at the borders of cells are clearly seen with aid of computer image treatment.

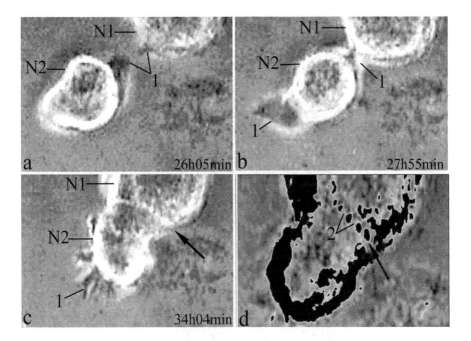

Figure 11. Formation of syncytial connection of two isolated neurons tissue culture.
a-c – approaching and fusion of neurons revealed in phase contrast; d – vacuole-like structures revealed with aid of computer Solarise effect (the same neurons as those in Fig. 1, c); 1 – nerve processes; N1, N2 – neurons; *arrows* – vacuole-like structures. Time – from the beginning of shooting. Obj. 40Ph, eyep. 10.

In semithin sections, we have managed to reveal along the contacting neuron edges the multiple outgrowths (the cytoplasmic feet) that are firmly adjacent to the "feet" of the neighbor cell. The paired feet are separated from each other by large vacuole-like "empty" structures that represent the local sharply enlarged fragments of the intercellular space (Fig. 12). The alternating bridges and the vacuole-like structures are clearly arranged along the cell borders and can serve a reliable orienteer of these borders under light microscope, especially at using the computer software ACDSee. It is these bridges that represent the cytoplasmic connections uniting the cytoplasm of adjacent cells. Here it is worth noting that the described vacuole-like local fusions are completely identical to such structures found at fusion of filopodia of growth cones [55, 56]. We have managed to detect the formed trinucleated neurons forming, in fact, the multinucleated symplast (Fig. 13).

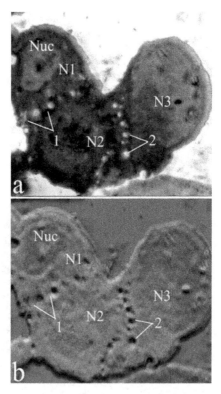

Figure 12. Light-optical signs of syncytial fusion of neurons in semithin sections.
a – additional staining with toluidine blue; b – vacuole-like structures revealed with aid of computer effect Emboss; 1 – cytoplasmic fusion bridges; 2 – vacuole-like enlargements of intercellular cleft; N1, N2 – adjacent neurons; Nuc – nuclei. Obj. 40Ph, eyep. 10.

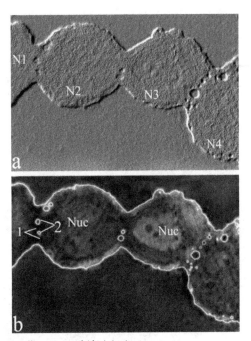

Figure 13. Chain of syncytially connected (draining) neurons.
a – computer effect of Emboss; b – computer effect Solarise; 1 – cytoplasmic bridges of fusion; 2 – vacuole-like enlargements of intercellular cleft; N1-N4 – fused neurons; Nuc – nucleus. Obj. 40Ph, eyep. 10.

With aid of electron microscope, indeed, it can be showing that bridges serve loci of fusion of two cells (Fig. 14). Although in some electron pictures the bridges of two contacting neurons can be separated with their external membranes (Fig. 14, b), the majority of membranes separating the cytoplasm of neighbor cells in the foot area turn out to be destroyed (Fig. 14, c, d). Instead of the external cell membranes separating the neuron cytoplasm, there are revealed only their short residual fragments that locally preserve the intercellular cleft, about 20 nm in width. In other places the neuroplasms of adjacent cells pass directly into each other (Fig. 15). Actually the cells turn out to be fused.

Thus, in these experiments, we were the first to manage modeling the syncytial connection between neurons *in vitro*, to prove their fusion, and thereby to confirm the principal similarity of neurons with other, non-nervous cells in the subject of intercellular interactions. Besides, results of these experiments, in our opinion, answer the chief question of the discussion about the principal possibility or impossibility of the syncytial connection of neurons. Not the initial small membrane pores and perforations have been demonstrated, but almost complete destruction of membranes of paired neurons and fusion of their cytoplasm have been shown. Fusion of neurons has been shown to occur not simultaneous on the entire contacting surface.

Initially, there are formed the fusion bridges, the local sites where membranes of contacts are destroyed and intercellular clefts between them are sharply enlarged in places and remind vacuoles (Fig. 16). This is the first absolute parameter of fusion.

The second absolute parameter of fusion of neurons is proven by McCarthy et al. [51]. It consists in that the adjacent cells first have the shape of the figure eight with the acute angle of waist. Then such structure gradually loses the shape of zero, i.e., strives for the shape of ball. The authors showed experimentally, with aid of microelectrode technique, that if the waist angle between paired contacting neurons exceeded 125°, these neurons had already fused (Fig. 16, *c, d*).

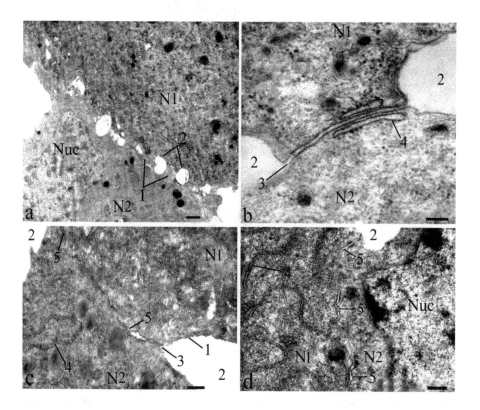

Figure 14. Borders of fused neurons.

a – multiple formations of cytoplasmic bridges of fusion and vacuole-like enlargements of intercellular cleft between two contacting neurons; b – preserved external cell membranes on the border of cytoplasmic bridges of two neurons; c, d – variants of destroyed borders between two fused neurons in the area of cytoplasmic bridge; 1 – cytoplasmic bridges of contacting neurons; 2 – vacuole-like enlargements of intercellular cleft; 3 – intercellular cleft; 4 – cistern of endoplasmic network; 5 – residual fragments of destroyed membranes on the border of two neurons; N1, N2 – adjacent neurons; Nuc – nucleus. Bar: a – 0.05 nm; b-d – 0.1 nm.

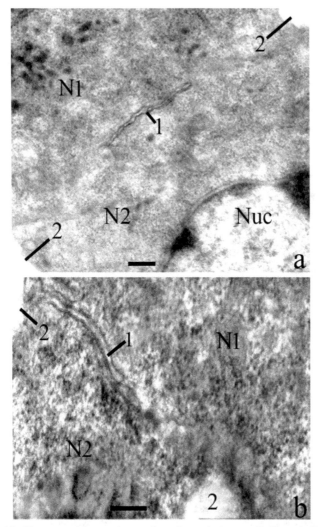

Figure 15. Residual fragments of destroyed external membranes in the area of bridges of adjacent neurons. a, b – variants of fragments; 1 – remnants of adjacent membranes with rounded ends; 2 – vacuole-like enlargements at the site of intercellular cleft. N1, N2 – adjacent neurons; Nuc – nucleus. Bar – 0.1 nm.

The data obtained in model experiments in tissue culture about the syncytial fusion of neuronal bodies are leading us to another, seemingly independent problem of mechanism of formation of binucleated and polynucleated cells [12].

The doubtless proves of formation of neurosyncytia have also revealed in tissue culture with aid of the phase-contrast microscopy of the structural neuron kinetics.

1. Syncytial connection between nerve processes is found in the case that nerve processes of some neurons, which contact with branches of another neuron do not die if the body of this neuron dies or is amputated. By the law of Wallerian degeneration the process separated from the cell body is to die necessarily. But if it does not degenerate, it means that it is connected cytoplasmically with the other neuron body.

2. The cytoplasmic syncytial connection of neurons can also be considered in the case that cytoplasmic varicosities are translocated from branch of one neuron to the branch of the other, contacting neuron.

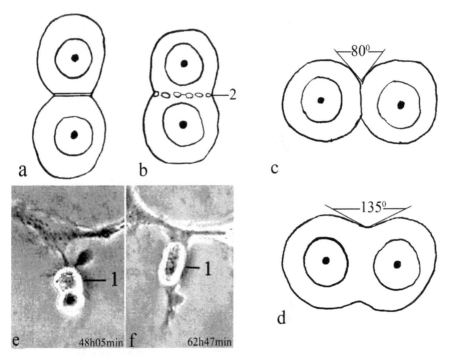

Figure 16. Light optical morphological parameters of syncytial fusion of neurons.
a, b – formation and visualization of fusion bridges and of bordering vacuole-like structures, rounding of the total structure of two fused cells; c, d – an increase of the waist angle of two fused cells higher than 135°; e, f – rounding of fused neurons; 1 – neuronal body; 2 – boundary vesicle-like structures; e, f – phase contrast. Obj. 20Ph, eyep. 10.

3. In semithin sections and under electron microscope, parameters of fused neurons serve the contacting "feet" of adjacent cells forming the fusion bridges, in which the residual fragments of destroyed boundary membranes of adjacent neurons are detected. The same criteria of syncytial connection serve the vacuole-like structures, significant local enlargements of intercellular clefts located between the fusion bridges. The local vacuole-like enlargements are also seen at phase-contrast microscopy on living fused neurons.

4. As seen with aid of a microelectrode leading and of recording of membrane permeability to stains with different molecular masses, an increase of the waist angle of attached neurons higher than 125° corresponds to conversion of gap junction pores into syncytial perforation and to fusion of neurons [51].
5. Since we have proven the binucleated neurons to be formed by the cytoplasmic syncytial fusion, their presence also is the absolute parameter of the existence of the syncytial connection in the nervous system.

7. History of study of binucleated neurons

The binucleated nerve cells were first revealed by I. Remak [57]. He was a pupil of the famous physiologist I. Müller who was the fist to describe the binucleated non-nerve cells. Since then the binucleated neurons were studied hundred times by numerous well-known histologists, such as Ranvier, Schtër, Dogiel, Schpielmayer, Bielschowski, Këliker, Alzheimer, and others [58, 59]. Their description had been included into previous textbooks and manuals past [60-62].

This problem has also been studied at present [63-65]. The binucleated and polynucleated neurons have been revealed in various central and peripheral parts of the nervous system [66, 67]: in large hemisphere cortex, hippocampus, globus pallidus, brainstem, cerebellum, pituitary, spinal cord, spinal ganglia, boundary sympathetic trunk, prevertebrate ganglia. The binucleated neurons were studied in human, monkey, horse, pig, sheep, rabbit, cat, dog, rat, mouse, guinea pig, frog, cod. In investigation of our Laboratory, they have been detected in intramural nerve plexuses of the normal cat gut (Fig. 17) and in dog [68] in toxicosis on the background of the portal vein (Fig. 18). More than a half of the authors studied the normal material.

In pathology there has also been described a great number of bi- and polynucleated neurons [69-74].

In the monograph of N.E. Yarygin and V.N. Yarygin [75] as well as in works of Ehlers [76], Shabadash and Zelikina [77] there are presented counts of binuclear neurons in norm. Their number in peripheral ganglia varies in different animals from 13 to 89%.

The issue of binucleated cells involves the problem of syncytial connection in the nervous system and the problem of non-divisibility of highly differentiated neurons. The most important is the issue of mechanism of formation of phenomenon of neuron binuclearity. The point is that from the very beginning of the studies and till now the appearance of binucleated neurons has been explained by division of mononucleated nerve cells. But since mitosis in differentiated neurons of adult animals is impossible, the hypothesis of the so-called amitosis was proposed, when the neuron nucleus is divided by direct perecording, without mitotic division figures. Since nobody has seen cytokinesis (the process of division of the cell body proper), there was formulated the additional hypothesis of the possibility of the incomplete amitosis restricted only by division of the nucleus. But division of the neuronal nucleus also was seen by nobody. All suggestions are based on voluntary selection of static fixed histological preparations.

This construction looks schematically like in Fig. 19, *a*. But the point is that in static histological preparations it is impossible to determine direction of vector of the process [78]. It is selected arbitrarily, based on the beforehand suggested hypothesis. Based on the same preparations it is also possible to model the process of the opposite direction: two neighbor cells are touched, form membrane contact, establish syncytial connection, and then are fused into one binucleated cell (Fig. 19, *b*).

However, for strange reason, almost all authors chose hypothesis of amitosis as mechanism of binuclearity and even do not discuss the possibility of mechanism of syncytial cytoplasmic fusion of neurons.

True, H. Apolant [79] mentions such possibility, but thinks it to be incompatible with physiological data. Indeed, the neuronal doctrine negates completely the possibility of interneuronal syncytium and, hence, does not accept the possibility of syncytial fusion. But in the culture of isolated neurons we have shown the binucleated neurons to be formed by syncytial fusion of usual mononucleated cells.

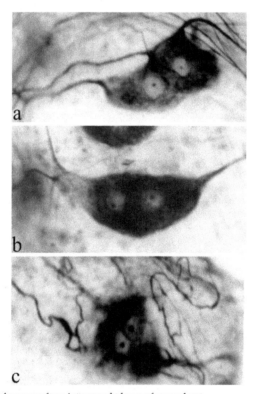

Figure 17. Binucleated neurons from intramural plexus of normal gut.
a – binucleated neuron with independent neurocytes; b – binucleated bipolar neuron with the generalized processes; c – a variant of binucleated neuron. Impregnation by Bielshowski-Gros. Obj. 40, eyep. 10.

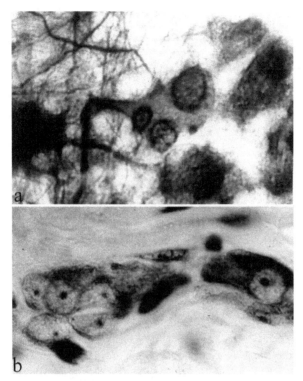

Figure 18. Polynucleated neurocytes in the node of submucosal plexus of the dog jejunum. The 15th day after the operation of the portal vein stenosing (by Chepur [68]). Procedures of Gomori in modification of Chilingaryan (a) and Einarson (b). Obj. 40, eyep. 10.

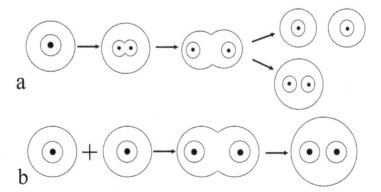

Figure 19. Schematic presentation of suggested mechanisms of formation of binucleated cells.
a – formation of binuclearity of neurons by amitotic cell division; b – formation of binuclearity by syncytial fusion of neurons.

8. Physiological proofs of interneuronal syncytial connection and of formation of binuceated neurons

The first electrophysiological study of neurons after their syncytial fusion with the neuroblasoma C 1300 neuroid cells seems to have been performed by A. Chalazonitis and coauthors [80]. The cells of the obtained hybrid line reminded the initial sympathetic neurons, were electroexcitable, sensitive to acetylcholine, and able to generate action potential. It is not necessary to prove that the hybrids are obtained by the cytoplasmic syncytial fusion. There also exist other examples that the presence of syncytial connection of neurons follows indirectly from physiological experiments.

In our work there remains one more uncovered point: whether there are physiological proofs of the existence of the syncytial interneuronal connection and what is the physiological effect of formation of syncytium in tissue culture.

In several experiments it was shown that such stains as Fura-2 and Calcium Green that are too large to overcome gap junction pores sometimes do penetrate across the boundary membranes. After injection into one cell they turned out to appear in the adjacent cell [81] as if the cleft contact would be converted into the syncytium. When studying the neuronal coupling only by the electrophysiological method, it is always difficult to distinguish the cleft contact from the electrical syncytial interconnection of neurons [82].

For the last few years there was intensively studied the role of interneuronal cleft contacts in hippocampus and cerebral cortex [83-85]. The electropermeable contacts were found to be the cause of formation of the high-frequency synchronization of the spike activity [63, 86]. It was shown that the "connexin-specific" blockators of cleft contacts: octanol, halotan, and carbenoxolen [87] cannot prevent completely effects of the presence of intercellular electrical connection. Suggestion of electrical connection of the "non-connexin type" appeared [88, 89]. This form of connection can be syncytium, as it is known that it is the syncytial perforations by their nature, which cannot be blocked by chemical agents.

It is also proven that the main protein providing intercellular connections in hippocampal cleft contact is connexin-36. It also serves the 100% marker of these contacts in vertebrates. However, in special studies of knockout animals deprived of connexin-36 and hence of cleft contacts [90, 91], the capability of hippocampus for synchronization of spikes as well as other effects are partly preserved [81, 92, 93]. Hence, apart from the cleft contact pores, hippocampus contains some intercellular connections between neurons, i.e., possibly the cytoplasmic syncytium.

Further. As long ago as in the 1990s a mysterious phenomenon was detected: the pseudorabies virus in the tissue culture was able to penetrate from neuron to neuron and to other cells by passing the intercellular medium [94-96]. This allowed suggesting formation of membranous perforations between cells. This virus was shown to be able to fuse neurons with the surrounding glia by forming binucleated cells and multinucleated hybrid symplasts [97].

Lastly, a work was published [51] whose authors infected culture of isolated neurons of rat sympathetic ganglion with the low pathogenic pseudorabies virus (PRV-151) and produced a massive formation of cleft contacts between adjacent neurons. The authors were able to demonstrate the ability of the dye with low molecular weight, Lucifer yellow (457 MW) to penetrate across the cleft contact membranous pores 9-12 h after infection of the culture. The dyes with the higher molecular weight could not be translocated from neuron to neuron. However, 24-26 h later, the dextran Texas red, the dye with the higher molecular weight (3000-40000 MW) began to penetrate from one neuron into the other one. This means that the cleft contact pores were transformed into syncytial cytoplasmic perforations; in other words, the neurons fused and acquired the dumbbell-like shape. The waist angle between fused neurons became more than 125° and several contacting neurons actually formed binucleated cells or the multinucleated symplast. Thus, with aid of the electrophysiological procedure there were shown the morphological equivalent and the physiological manifestation of the syncytial fusion of neurons.

It is important to emphasize that 18-20 h after infection of culture the incomplete electrical coupling (interneuronal electrical connection) was still observed, although a weak ability of the Lucifer yellow to penetrate from one cell into the other already was obvious. There were noted the incomplete coincidence of rhythms of adjacent cells, deletion of spikes or their conversion into spikelets (the noise-like spikes). But after 24-26 h the paired neurons became completely connected. They reproduced synchronously both spikes and spikelets, i.e., worked as the single binucleated cell.

Thus, these authors have managed to: first to obtain a model of the massive goal-seeking formation of gap junctions; second, with aid of microelectrode electrophysiological procedures to demonstrate conversion of small membranous pores of these contacts into syncytial perforations, i.e., to fuse neurons in the same way as we managed to do this in tissue culture without using virus.

In our work, by morphological methods it was also shown that syncytial perforations appeared in the place of gap junctions as a result of their perforation [25]. Hence, apart from the existing static concepts of structure of gap junctions there should be formulated a novel concept of transformation of cleft contacts from perforation of contacting membranes up to the complete fusion of neurons.

Some consequences of fusion of neuronal bodies can be revealed with aid of mathematical modeling [98, 99]. Due to development of the process of the neuronal body fusion and a rise of fusion pores between these neurons, the degree of synchronization of impulses of this pair of neurons increased. In the end of the fusion process the impulse activity of these neurons became practically synchronous. Besides, after fusion of two identical spherical neurons and at their transition to the shape of one sphere with preservation of the total cytoplasm volume the dikaryon surface will be reduced by approximately 20% as compared with that prior to fusion. Also reduced was the capacity of the cell body membrane. The newly fused nerve cell should be more sensitive to controlling signals of synapses.

Also attention is to be paid to data of study of neocortex of the 14-day rat embryos [12]. The syncytial cytoplasmic connections in these embryos were detected at the time when the developed chemical synapses still were absent. Besides, it is known that in pathology, when chemical synapses are damaged for the first time, a clear increase of the number and size of syncytial perforations is noted [35]. Possibly, in these cases, when the fine procession of information in chemical synapses is absent or attenuated, brain uses rough, but simple and stable form of impulse transduction between neurons with aid of syncytial pores and perforations. This primitive way is resistant to chemical actions and pessimal impulse activation and hence can have certain compensatory significance. Besides, the presence of the syncytial cytoplasmic connection between neurons allows the direct exchange with energetically important substances and proteins between the relatively preserved and the damaged neurons. This also can be considered as compensatory phenomenon.

In our neurohistological studies, on many objects there were recorded syncytial cytoplasmic interneuronal connections [100]; however, studies on living neurons in the tissue culture were necessary to demonstrate kinetics of their formation and transformation into fusion of cells and multinuclearity; it was necessary to develop the way of massive artificial obtaining in the tissue culture of interneuronal syncytial connections in experiment. Such way was developed [54]. We have managed to obtain fusion of two or several neurons, i.e., to model formation of bi- and multinucleated neurons and to confirm fusion of cells with aid of electron microscope.

The fact that syncytial perforations are formed on the basis of cleft contacts, while individual perforations are transformed with time into the total fusion of neurons with formation of binucleated cells allows formulating notion about the single process of readjustment of paired membranes. The process begins with formation of dotted and expanded membranous contacts, their fusion, formation of small pores and large perforations. Enlargement of perforations leads to fusion of neuronal bodies and formation of binucleated cells.

Acceptance of cytoplasmic syncytial connections in the nervous system as the morphological and physiological reality removes contradiction of the neuronal theory and cellular theory. Similarity of the main properties of all cells including their ability to form the cytoplasmic syncytium and fusion has been confirmed.

Author details

O.S. Sotnikov
Pavlov Institute of Physiology of the Russian Academy of Sciences, St. Petersburg, Russia

Acknowledgement

The work is supported by the Russian Foundation for Basic Research. The author is grateful to associates of the Laboratory A.A. Laktionova, N.M. Paramonova, T.V. Krasnova, E.A. Gendina and L.Z. Pevzner for assistance in creation of this Chapter.

9. References

[1] Gerlach J (1872) Von den Rückenmarke. In: Handbuch der Lehre von den Geweben des Menschen und der Thiere. 2; 30: In Tech. pp. 665-693.

[2] Ramon y Cajal S (1985) Recollection of My Life. Moscow. Meditsina: InTech. 283 pp. In Rus.

[3] Bullock TH, Bennett MVL, Johnston D (2005) The Neuron Doctrine, Redux. Science. 310: 791-793.

[4] Mazzarello P (2007) Net Without Nodes and Vice Versa, the Paradoxical Golgi-Cajal Story: a Reconciliation? Brain Res. Bull. 71: 344-346.

[5] Piccolino M (1988) Cajal and the Retina: a 100-year Retrospective. Trends Neurosci. 11: 521-525.

[6] Peters A (2007) Golgi, Cajal, and Fine Structure of the Nervous System. Brain Res. Rev. 55: 256-263.

[7] Torres-Fernández O (2006) The Golgi Silver Impregnation Method: Commemorating the Centennial of the Nobel Prize in Medicine (1906) Shared by Camillo Golgi and Santiago Ramón y Cajal. Biomedica. 26: 498-508.

[8] Robertis ED, Bennett E (1955) Electron Microscope Observations on Synaptic Vesicles in Synapses of the Retinal Rods and Cones. J. Biophys. Biochem. Cytol. 1: 47-58.

[9] Palay SL (1958) The Morphology of Synapses in the Central Nervous System. Exp. Cell Res. 5: 275-293.

[10] Makovetskii AN (1940-1941) Sophists. Baku. In Tech. 1-2: 48 pp. In Rus.

[11] Ramon y Cajal S (1954) Neuron Theory or Reticular Theory? Objective Evidence of the Anatomical Unity of Nerve Cells. Madrid. Instituto "Ramon y Cajal": In Tech. 8: 144 p.

[12] Sotnikov OS, Frumkina LE, Novakovskaya SA, Bogolepov NN (2011) Fusion of Brain Neurons in Rat Embryos. Morfology. 139: 18-21. In Rus.

[13] Dogiel AS (1893) To the Question of Relation of Nerve Cells to Each Other. In: Histological Studies. I, Tomsk, P.I. Mokushin's Typo-Lithography, 23 pp. In Rus.

[14] Jabonero V. (1956) Studien über die Synapsen des Periferen Vegetativen Nervensystems. III. Das Distale Nervosa Synzytium und die Plexiforme Synapse auf Distanz. Z. Mikr.-anat. Forsch. 62: 407-451.

[15] Amzica F (2003) Physiology of Sleep and Wakefulness as it Relates to the Physiology of Epilepsy. J. Clin. Neurophysiol. 19: 488-503.

[16] Young JZ (1939) Fused Neurones Synaptic Contacts in the Giant Nerve Fibres of Cephalopods. Philosoph. Transact. Roy. Soc. London: Ser B, Biol. Sci. 229: 465-503.

[17] Maximov AA, Bloom W (1938) A Text–Book of History, W.B. Saunders Company, Philadelphia, 380 pp.

[18] Sotnikov OS, Kamardin NN, Rybakova GI, Solovieva IA (2009). Cytoplasmic Syncytial Interneuronal Connection in Molluscs. J. Evol. Biochem. Physiol. 45: 223-232. In Rus.

[19] Günter J (1975) Neuronal Syncytia in the Giant Fibres of Earthworms. J. Neurocytol. 4: 55-62.

[20] Hagiwara S, Morita H, Naka K (1964) Transmission Through Distributed Synapses Between the Giant Axons of a Sabellid Worm. Comp. Biochem. Physiol. 13: 453-460.

[21] Nicol JAC (1948) Giant Axons of *Eudistylia vancouveri* (Kinberg) Transact. Roy. Soc. Canada, XIII, (III): 107-124.

[22] Nicol JAC, Young JZ (1946) Giant Nerve Fibre of *Myxicola infundibulum* (Grube). Nature. 158: 167-168.

[23] Young, JZ (1936) Structure of Nerve Fibres and Synapses in Some Invertebrates. Cold Spring Harbor. Symp. Quant. Biol. 4: 1-6.

[24] Young JZ (1938) The Functioning of the Giant Nerve Fibres of the Squid. J. Exper. Biol. 85: 170-185.

[25] Guillery RW (2007) Relating the Neuron Doctrine to the Cell Theory. Should Contemporary Knowledge Change Our View of the Neuron Doctrine? Brain Res. Rev. 55: 411-421.

[26] Birse SC, Bittner GD (1976) Regeneration of Giant Axons in Earthworms. Brain Res. 113: 575-581.

[27] Derimer SA, Elliot EJ, Macagno ER, Muller KJ (1983) Morphological Evidence That Regeneration Axons Can Fuse with Axon Segment. Brain Res. 272: 157-161.

[28] Hoy RR, Bittner GD, Kennedy D (1977) Regeneration in Crustatian Motoneurons Evidence for Axon Fusion. Science. 156: 251-252.

[29] Sotnikov OS, Frumkina LE, Maiorov VN, Paramonova NM, Laktionova AA, Bogolepov NN (2012) Rehabilitation of Interneuronal Syncytial Connection in the Nervous System. Pacific Med. J. 2: 75-83. In Rus.

[30] Sotnikov OS, Lagutenko YuP, Malashko VV, Gusova BA, Podolskaya LA (1994) Kinetics of Avesicular Interneuronal Membranous Contacts. III Congr. Anatom. Histolog. Embryol. Ross. Federat. Minzdrav, Tyumen, 189. In. Rus.

[31] Sotnikov OS, Kamardin NN, Rybakova GI (2009) Cytoplasmic Syncytial Interneuronal Connection in Molluscs. J. Evol. Biochem. Physiol. 45: 223-232. In Rus.

[32] Sotnikov OS, Malashko VV, Rybakova GI (2006) Fusion of Nerve Fibers. Proceed. Biol. Sci. 410: 361-363. In Rus.

[33] Sotnikov OS, Fomichev NI, Laktionova AA, Archakova LI (2010) Glio-neuronal and Glio-glial Syncytial Cytoplasmic Connections in Peripheral Nerve Trunks of the Crawfish *Astacus leptodactylus*. J. Evol. Biochem. Physiol. 46: 429-434. In Rus.

[34] Archakova LI, Sotnikov OS, Novakovskaya SA (2010) Syncytial Cytoplasmic Ganglion Cells in Adult Cats. Neurosci. Behav. Physiol. 40: 447-450.

[35] Paramonova NM, Sotnikov OS, Krasnova TV (2010) Interneuronal Membrane Contacts and Syncytial Perforations in CA2 Hippocampal Area After Brain Trauma. Bull Exp. Biol. Med. 150: 100-103.

[36] Sotnikov OS,, Paramonova NM (2010) Cytoplasmic Syncytial Connection – One of Three Forms of Interneuronal Connection. Usp. Physiol. Science 41(1): 45-57. In Rus.

[37] Sotnikov OS, Novakovskaya SA, Solovieva IA (2011) Syncytial Perforations of Human Embryo Neuronal Membranes. Russ. J. Devel. Biol. 42(1): 48-52.

[38] Kostenko MA, Sotnikov OS, Chistyakova IA (1999) Methods and Methodological Approaches to Studies of Isolated Neurons of Brain from Adult Animals (*Lymnaea stagnalis*) in Tissue Culture. Neurosci. Behav. Physiol. 29: 455-459.

[39] Petrova IO (1976) Proteolytic Enzymes of Actinomycetes. Moscow. Nauka: In Tech. 26 pp. In Rus.

[40] Leprince P, Register B, Delree P (1991) Proteases et Inhibiteurs de Proteases: Implications Multiples Dans le Development et le Vieillissement Cerebral. Rev. ONO. 12-13: 30-38.

[41] Scher M, Aidoo R, Chung P (1991) Plasminogen Activator is Regulated in Peripheral Nerve After Injury and During Regeneration. J. Cell Biol. 115(2): 117.

[42] Kostenko MA (1972) Isolation of Single Nerve Cells from Brain of the Mollusc *Lymnaea stagnalis* for Their Further Culturing *in vitro*. Cytologia. 14(10): 1274-1279. In Rus.

[43] Kostyuk PG., Kryshtal OA (1981) Mechanisms of Electrical Excitability of the Nerve Cell. Moscow. Nauka: In Tech. 108 pp.

[44] Akaike N, Nishi K, Oyama Y (1983) Characteristics of Manganese Current and Its Comparison With Current Carried by Other Divalent Cations in Snail Some Membrane. J. Membr. Biol. 76: 289-297.

[45] Zhang W, Lin GJ, Takeuchi H (1996) Effect of L-Glutamic Acid and Its Agonists on Snail Neurons. Gen. Pharmacol. 27(3): 487-497.

[46] Sotnikov OS, Chalisova NI (1988) Supravital Methods of Morphological Investigations. In: Manual of Culturing of Nervous Tissue. Moscow. Nauka: In Tech. pp. 79-84. In Rus.

[47] Fikhte BA, Gurevich GA (1988) Deintegrators of Cells. Moscow. Nauka: In Tech. 185 pp. In Rus.

[48] Sotnikov OS (1985) Dynamics of Structure of Living Neuron. Leningrad. Nauka: In Tech. 159 pp. In Rus.

[49] Sotnikov OS, Malashko VV, Rybakova GI (2008) Syncytial Coupling of Neurons in Tissue Culture and Early Ontogenesis. Neurosci. Behav. Physiol. 38: 223-331.

[50] Waller A (1850) Experiments on the Section of the Glossopharyngeal and Hypoglossal Nerves of the Frog and Observations of the Alterations Produced Thereby in the Structure of Their Primitive Fibres. Philos. Trans. 140: 423-429.

[51] McCarthy KM, Tank DW, Ehquist LW (2009) Pseudorabies Virus Infection Alters Neuronal Activity and Connectivity *in vitro*. PLoS Pathol. 5(10): 1-20.

[52] Efrussi B. (1976) Hybridization of Somatic Cells. Moscow. Mir: In Tech. 189 pp. In Rus.

[53] Lentz BR (2007) PEG as a tool to gain insight into memrane fusion. Eur. Biophys. J. 36(4-5):315-326.

[54] Sotnikov OS, Laktionova AA, Paramonova NM (2011) Way of Modeling of Syncytial Connections Between Cells *in vitro*. Patent RU 2010 114371. Decision of Granting the Patent 2011.05.30. In Rus.

[55] Luduena MA, Wessels NK (1973) Cell Locomotion, Nerve Elongation and Microfilaments. Dev. Biol. 30: 427-440.

[56] Spooner BS, Luduena MA, Wessells NK (1974) Membrane Fusion in the Growth Cone-Microspike Region of Embryonic Nerve Cells Undergoing Axon Elongation in Cell Culture. Tissue Cell. 6(3): 399-409.

[57] Remak R (1837) Weitere Mikroscopishe Untersuchungen über die Premitivfasern der Nervensystems der Wirbelthiere. Notizen aus dem Bau der Natur und Helkinde. 3(47): 36-38.

[58] Das GD (1977) Binucleated Neurons in the Central Nervous System of the Laboratory Animals. Experientia. 33(9): 1179-1180.

[59] Mair H, Budka H, Lassman H (1989) Vacuolar Myelopathy with Multinucleated Giant Cells in the Acquired Immune Deficiency Syndrome (AIDS). Light and Electron Microscopic Distribution of Human Immunodeficiency Virus (HIV). Acta Neuropathol. 78(5): 497-503.

[60] Spielmeyer W (1922) Histopathologie des Nervensystems. Berlin. Verlag von J. Springer: In Tech. 498 pp.

[61] Botar J (1966) The Autonomic Nervous System. Budapest. Akademiai.: In Tech. 357 pp.

[62] Ermokhin PN (1969) Histology of the Central Nervous System. Moscow. Meditsina: In Tech. 243 pp. in Rus.

[63] Maier N, Nimmrich V, Draguhn A (2003) Cellular and Network Mechanisms Underlying Spontaneous Sharp Wave – Ripple Complexes in Mouse Hippocampal Slices. J. Physiol. 550(3): 873-887.

[64] Kawataki T, Sato E, Sato T (2010) Anaplastic Ganglioglioma With Malignant Features In Both Neuronal And Glial Components – Case Report. Neurol. Med. Chir. 50(3): 228-231.

[65] DiLorenzo DJ, Jankovic J, Simpson RK (2010) Long-Term Deep Brain Stimulation for Essential Tremor: 12-year Clinicopathologic Follow-up. Mov. Disord. 25(2): 232-238.

[66] Tambuyzer BR, Nouwen EJ (2005) Inhibition of Microglia Multinucleated Giant Cell Formation and Induction of Differentiation by GM-CSF Using a Porcine *in vitro* Model. Cytokine. 31(4): 270-279.

[67] Mizuguchi M (2007) Light Microscopic Analysis of Cellular Networks in the Pineal Gland of the Golden Hamster as Revealed by Methylene Blue Labeling. Congenit. Anom. 47(1): 2-8.

[68] Chepur SV (2002) Morphofunctional Characteristics of the Nervous System Structures in Norm and Regularities of Their Changes in Hepatic Encephalopathy. Doctorate Dissert. Pavlov Institute of Physiology: St. Petersburg. 390 pp. In Rus.

[69] Yamanouchi H, Jay V, Rutka JT (1997) Evidence of Abnormal Differentiation in Giant Cells of Tuberous Sclerosis. Pediatr. Neurol. 17(1): 49-53.

[70] Zhu X, Siedlak SL, Wang Y (2008) Neuronal Binucleation in Alzheimer Disease Chihppocampus. Neuropathol. Appl. Neurobiol. 34(4): 457-465.

[71] Paltsyn AA, Kolokolchikova EG, Kostantinova NB (2008) Formation of Heterokaryons as the Way of Regeneration of Neurons at Postischemical Damage of Rat Brain Cortex. Bull. Exper. Biol. Med. 146(10): 467-470. In Rus.

[72] Hirohata S (2008) Histopathology of Central Nervous System Lesions in Behçet's Disease. J. Neurol. Sci. 267(1-2): 41-47.

[73] Konstantinova NB (2010) Role of Cell Fusion in Reparative Regeneration of the Brain Cortex. Candidate Dissertation, Research Inst. Gener. Pathol. Pathophysiol. Ross. Acad. Med. Sci. Moscow: 24 pp. In Rus.

[74] Samosudova NV, Reutov VG, Larionova NP (2011) Fusion of Cells-Granules of Frog Cerebellum at the Toxic Effect of Glutamate and a NO-Generating Compound, Morphologia. 140(4): 13-17. In Rus.

[75] Yarygin NE, Yarygin VN (1973) Pathological Adaptive Neuron Changes. Moscow. Meditsina: In Tech. 175 pp. In Rus.

[76] Ehlers P (1951) Über Altersveränderungen an Grenzstrang-Ganglien vom Meerschweinchen. Anat. Anz.. Bd. 98, H. ½. S. 24-34.

[77] Shabadash AL, Zelikina TI (1968) Cytochemistry of Experimental Karyotomy in Nevrocytes of Mammalian Autonomic Ganglia. Proc. Acad. Sci. USSR. 183(4): 944-947. In Rus.

[78] Wang XJ, Li QP (2007) The Roles of Mesenchymal Stem Cells (MSCs) Therapy in Ischemic Heart Diseases. Biochem. Biophys. Res. Commun. 359(2): 189-193.

[79] Apolant H (1896) Über die Sympathischen Ganglienzellen der Nager. Arch. Mikroskop. Anat. Entwick.. Bd. 47: S. 461-471.

[80] Chalazonitis A, Greene LA, Shain W (1975) Excitability and Chemosensitivity Properties of a Somatic Cell Hybrid Between Mouse Neurol Laltoma and Sympathetic Ganglion Cells. Exp. Cell Res. 96: 225-238.

[81] Buzsáki G (2001) Electrical Wiring of the Oscillating Brain. Neuron. 31(3): 342-344.

[82] Thompson RJ, Zhou N, MacVicar BA (2006) Ischemia Opens Neuronal Gap Junction Hemichannels. Science. 312(5775): 924-927.

[83] Venance L, Rozov A, Blatow M (2000) Connexin Expression in Electrically Coupled Postnatal Rat Brain Neurons., Proc. Natl. Acad. Sci. USA. 97(18): 10260-10265.

[84] Bruzzone R, Hormuzdi SG, Barbe MT (2003) Pannexins, a Family of Gap Junction Proteins Expressed in Brain. Proc. Natl. Acad. Sci. 100(23): 13644-13649.

[85] Hamzei-Sichani F, Kamasawa N, Janssen WG (2007) Gap Junctions on Hippocampal Mossy Fiber Axons Demonstrated by Thin-Section Electron Microscopy and Freeze Fracture Replica Immunogold Labeling. Proc. Natl. Acad. Sci. 104(30): 12548-12553.

[86] Schmitz D, Schuchmann S, Fisahn A (2001) Axo-Axonal Coupling a Novel Mechanism for Ultrafast Neuronal Communication. Neuron. 31(5): 831-840.

[87] Draguhn A, Traub RD, Bibbig A (2000) Ripple (Approximately 200-Hz) Oscillations in Temporal Structures. J. Clin. Neurophysiol. 17(4): 361-376.

[88] Ylinen A, Bragin A, Nádasdy Z (1995) Sharp Wave-Associated High-Frequency Oscillation (200 Hz) in the Intact Hippocampus: Network and Intracellular Mechanisms. J. Neurosci. 15(1): 30-46.

[89] Stebbings LA, Todman MG, Phillips R (2002) Gap Junctions in Drosophila: Developmental Expression of the Entire Innexin Gene Family. Mech. Dev. 113(2): 197-205.

[90] Kistler WM, De Jeu MT, Elgersma Y (2002) Analysis of Cx36 Knock-out Does not Support Tenet that Olivary Gap Junctions are Required for Complex Spike Synchronization and Normal Motor Performance. Ann. NY Acad. Sci. 978: 391-404.

[91] Buhl DL, Harris KD, Hormuzdi SG (2003) Selective Impairment of Hippocampal Gamma Oscillations in Connexin-36 Knock-out Mouse in vivo. J. Neurosci. 23(3): 1013-1018.

[92] Deans MR, Gibson JR, Selitto C (2001) Synchronous Activity of Inhibitory Networks in Neocortex Requires Electrical Synapses Containing. Connexin 36. Neuron. 31: 477-485.

[93] Traub RD, Pais J, Bibbig A (2002) Containing Roles of Axonal (Pyramidal Cell) and Dendritic (Interneuron) Electrical Coupling in the Generation of Neuronal Network Oscillations. Proc. Natl. Acad. Sci. USA. 100(3): 1370-1374.

[94] Card JP, Rinaman L, Schwaber JS (1990) Neurotrophic Properties of Preudorabies Virus: Uptake and Transneuronal Passage in the Rat Central Nervous System. J. Neurosci. 10(6): 1974-1994.

[95] Ch'ng TH, Enquist LW (2005) Neuron-to-Cell Spread of Pseudorabies Virus in a Compartmented Neuronal Culture System. J. Virol. 79(17): 10875-10889.

[96] Ch'ng TH, Spear PG, Struyf F (2007) Glycoprotein D-Independent Spread of Pseudorabies Virus Infection in Cultured Peripheral Nervous System Neurons in a Compartmented System. J. Virol. 81(19): 10742-10757.

[97] Reichelt M, Zerboni L, Arvin AM (2008) Mechanisms of Varicella-Zoster Virus Neuropathogenesis in Human Dorsal Root Ganglia. J. Virol. 82(8): 3971-3983.

[98] Pokrovskii AN, Sotnikov O.S. (2011) Mathematical Modeling of Consequences of Fusion of Neuronal Bodies. Proc. Intern. Conf. Voronezh. Izdatelsko-Poligraficheskii Tsentr VGU: In Tech. 233-234. In Rus.

[99] Pokrovsky AN, Sotnikov O.S. (2008) Difficulties in Mathematical Modelling of Control Processes in One-Type Neuron Populations. II International Conference on Inductive Modelling. Kyiv. NAS of Ukraine: In Tech. 112-115.

[100] Sotnikov OS (2008) Statics and Structural Kinetics of Living Asynaptical Dendrites. St. Petersburg. Nauka: In Tech. 397 pp. In Rus.

Chapter 7

Tissue Development and Mechanical Property in the Regenerated-Cartilage Tissue

Seiji Omata, Yoshinori Sawae and Teruo Murakami

Additional information is available at the end of the chapter

1. Introduction

Vertebrates have multiple synovial joints, and these joints in humans enable a number of activities of daily life. In particular, the hip, knee, shoulder, elbow, and ankle are all synovial joints with synovium in which the secretes synovial fluid. The articular surface is covered with smooth hyaline cartilage, which forms the part of the skeletal system that is notably different from mineralized bone in both function and histological composition [1]. The formation of animal cartilage from mesenchyme occurs in numerous areas of the embryo, such as the skull, limbs, and spine. This tissue contains chondrocytes which synthesize and maintain extracellular matrix (ECM), which is composed of a dense network of collagen molecules and proteoglycans [2]. There are three types of cartilage; fibrocartilage, elastic cartilage, and hyaline cartilage. Fibrocartilage contains regions of organized fibrous tissue, containing type I collagen in addition to the normal type II collagen. It is found in the annulus fibrosus of the intervertebral discs, meniscus and temporomandibular joints. Elastic cartilage contains additional elastin fibers and is a type of cartilage present in the ear, larynx, and epiglottis. Finally, hyaline cartilage coats the articular surfaces of bone epiphyses and is composed of individual chondrocytes bound together by the ECM. The major constituent of the hyaline cartilage is held in place by proteoglycan (about 10 % of wet weight) and type II collagen (10–20 % of wet weight), which forms a meshwork with high tensile strength [3–5].

Steric and electrostatic interactions of these ECM molecules in the hyaline cartilage occur between the cationic collagen fibers and the anionic proteoglycans to provide a highly charge in a neutral pH environment. In particular, since the proteoglycan is highly negatively charged, it is not only cross-liked between collagen, but also combined with many water molecules, leading to high extracellular osmolality within the cartilage. Consequently, cartilage is as formed as a soft-tissue-bearing element with high viscoelasticity [6]. The mechanical function of the hyaline cartilage is joint lubrication and shock absorption, and articular cartilage has a very low friction coefficient (0.02) [7] with smoother movement than most modern artificial joint replacements.

Although adult articular cartilage is a remarkable load bearing system, it lacks the ability to repair itself under conditions of wear and tear or traumatic injury, leading to osteoarthritis (OA). This lack of self-maintenance is due to its avascular, aneural, alymphatic and almost nonimmunogenic properties, as well as its nourishment entirely via diffusion from synovial fluid. Diseases of the cartilage are a major health problems, especially in industrialized countries with long life expectancy. At present, there is no established therapy of cell-assisted tissue regeneration for sufficiently reliable and durable replacement of damaged articular cartilage [8–11]. In recent years, however, tissue engineering has shown promise toward the treatment of OA, enabling researchers to produce functional replacements for diseased cartilage [12, 13]. Developments in therapeutic strategies for damaged cartilage treatment have increasingly focused on the promising technology of cell-assisted repair and proposing the use of autologous chondrocytes or other cell types to regenerate articular cartilage [10, 11].

Little knowledge is available for establishing a suitable design strategy for reconstructing cartilage by tissue engineering to match the mechanical properties of natural tissue. It is therefore necessary to understand the relationship between degree of development of the ECM network and the macroscopic mechanical properties of the cultured construct for producing regenerated cartilage. Because the ECM meshwork is formed chiefly by type II collagen, growth of the collagen network likely plays the most important role in the mechanical characteristics of the overall three-dimensional construct (e.g., Young's modulus and compressive strength). If ECM network does not interconnect among chondrocytes in the cell-scaffold material, it is no relevant to entire mechanical property of the construct. In such a case, this mechanical property will be the same as that of the scaffold material without chondrocytes. This means that less mechanical property increases if the ECM synthesizes a lot. We therefore investigated the relationship between development of the ECM network and the macroscopic mechanical property of the cultured construct using vitamin C (VC) to control collagen synthesis [14].

VC is an important water-soluble antioxidant and enzyme cofactor of collagen synthesis in plants and animals [15] and the importance of VC in collagen synthesis is well-known as described below [16–18]. Procollagen molecules, which are collagen precursors, undergo multiple steps of post-translational modification. After translocation of the growing polypeptide chains of procollagens into the rough endoplasmic reticulum, hydroxylation of proline residues is catalyzed by prolyl 4-hydroxylase (P4H). This catalytic reaction requires ferrous ions, 2-oxoglutarate, molecular oxygen, and ascorbic acid (AsA), a kind of VC, as cofactors. The hydroxylation of proline residues increases the stability of the triple helix and is a key element in its folding. P4H requires an unfolded chain as a substrate. The C-propeptides have an essential function in the assembly of the three α-chains into trimeric collagen monomers. The formation of triple helices starts from the alignment of the C-terminal domains of the three α-chains and proceeds to the N-terminal. Premature association between procollagen is thought to be prevented by heat shock protein 47 [18] and by collagen-modifying enzymes until biosynthesis of the individual chain is complete. In the collagen-secretion process, proline hydroxylation is caused by oxidation of P4H-bound ferrous iron, which must be reduced to P4H-bound ferric iron by AsA to reinvigorate P4H activity for maintaining the proline hydroxylation process. Thus, the collagen biosynthesis needs AsA to reduce oxidized P4H-bound iron.

Many studies have reported the relationship between collagen content in the tissue and mechanical properties [19–21] and the effects of VC on several cell functions (e.g., cytotoxicity and redox) [15, 22–25]; however, little knowledge is available on the development of the ECM network in relation to the mechanical properties of the cultured construct. In this chapter, we demonstrate the influence of collagen network growth on the macroscopic mechanical properties of regenerated cartilage using a chondrocyte-agarose construct, and we briefly explain how expansion of the network within a tissue affects to the mechanical properties. Here we used two types of VC: AsA, the acidic form; and ascorbic acid 2-phosphate (A2P), the non-acidic form. In addition, after applying uniaxial compressive strain to the tissue model using a purpose-built bioreactor, we described the different ways of developing the ECM in the construct by dosing the culture medium with AsA or A2P.

2. Materials & methods

2.1. Sample preparation and culturing

Primary bovine chondrocytes were isolated from metacarpophalangeal joints of steers purchased from a meat center in Fukuoka city, Japan using a sequential enzyme digestion method [26]. Full-thickness articular cartilage tissue was harvested from the proximal articular surface of the metatarsal bone and finely diced with a scalpel. The finely diced cartilage was enzymatically digested with 25 unit/mL protease solution (P8811, Sigma, St Louis, MO) for 3 hours, and subsequently with 200 unit/mL collagenase solution (C7657, Sigma) for 18 hours at 37 degree C. Both enzyme solutions were prepared in sterile tissue culture medium, consisting of Dulbecco's modified Eagle's medium (DMEM; D5921, Sigma) supplemented with 20 v/v% Fetal Bovine Serum (FBS; 10437-028, Gibco, CA), 2 mM L-glutamine (G7513, Wako Pure Chemical Industries, Ltd., Osaka, Japan), 100 unit/mL penicillin, 100 μg/mL streptomycin, 0.25 μg/mL amphotericin B (161-23181, Wako), 20 mM hepes (H0887, Sigma), and 0.85 mM L-ascorbic acid (AsA; A5960, Sigma). The supernatant of the resultant solution was centrifuged to separate chondrocytes at $40 \times g$ for 5 min. The resultant cell pellet was washed twice with fresh culture medium, and cell number and cell viability were determined by trypan blue assay. In this study, Sigma Type VII agarose (A6560, Sigma) was used to prepare a chondrocyte-agarose construct. The agarose powder was dissolved in Earle's balanced salts solution (EBSS; Sigma) at twice the required final concentration (1 w/v%) and mixed with an equal volume of the cell suspension to yield the desired agarose concentration with a final cell density of 1×10^7 cells/mL. The molten cell-agarose solution was poured into an acrylic mold and quenched to gel at 4 degree C for 30 min to create cylindrical constructs with a diameter of 4 mm and a height of 2.5 mm. The resultant cell-agarose constructs were placed into a 24-well culture plate and subsequently cultured with 1 mL culture medium in a humidified tissue culture incubator controlled at 37 degree C and 5 % CO_2. The culture medium was exchanged every two days. Several culture media with different VC concentrations, from 0.64 to 6.4 pmol/10^9 cells were prepared and used to evaluate the effect of AsA and A2P Salt (Wako) concentrations. The culture medium without VC was also used as a control. Media are denoted abbreviated as AsA(0.64), AsA(2.2), AsA(3.2), A2P(3.2), and A2P(6.4) hereafter in this paper.

2.2. Mechanical stimulation

After 1 day of free-swelling culture, the constructs in the experiment group were subjected to uniaxial compression within a purpose-built bioreactor system as shown in Fig. 1. The system enables the application of strain independently in each vertical and horizontal direction to

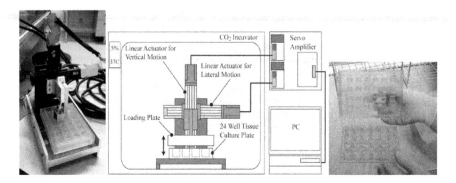

Figure 1. Photograph (left) and Schematic drawing (center) of the mechanical loading system mounted within tissue culture incubator and appearance of the loading platen with 22 plungers coupled with a 24 well culture plate (right).

individual constructs using a 24-well plate in a commercially available incubator. Movements was controlled with respective linear variable displacement transducers and linear guide actuators. Strain was applied to the individual constructs through a loading plate, which was attached to the actuator via a jig. Uniaxial cyclic compression up to a maximum amplitude of 15 % was applied with a triangular waveform at a frequency of 1 Hz for 6 hours and subsequently off-loaded with the platen resting on each construct for the subsequent 18 hours. Control constructs were cultured in contrast with both upper- and lower-platens to diffusion through its sides alone.

2.3. Mechanical testing

To examine the influence of chemical and physical stimulation on the mechanical properties of the constructs, cell-agarose constructs were subjected to unconfined compression while immersed in culture medium at room temperature. Individual constructs were tested after culture periods of 1, 8, 15, and 22 days. Mechanical tests were performed with an impermeable stainless steel plunger at a strain rate of 0.5 mm/min up to a strain of 10%, while the load was recorded with a 10-N load sensor, as shown in Fig. 2. The tangent modulus of the construct was calculated from the slope of a straight-line approximation of the stress-strain curve with a range of 0–15 % strain using the least-square method.

2.4. Immunohistology

Separate constructs were used for trichrome immunofluorescence observation to examine the morphological characteristics of the elaborated ECMs, in particular type I collagen, type II collagen, and chondroitin sulphate. After the prescribed culture periods, representative constructs were cut into slices with a thickness of approximately 1 mm using a knife. The slices were washed in Ca^{2+} and Mg^{2+} free phosphate-buffered saline (PBS(–)) and subsequently incubated in PBS(–) + 1 w/v% bovine serum albumin (BSA; Wako) for 30 min at 37 degree C. These slices were incubated in PBS(–) containing the three monoclonal antibodies (bovine IgG1 isotype anti-type I collagen, Funakoshi, Tokyo, Japan; embryonic chicken IgG2a isotype anti-type II collagen, Funakoshi; mouse IgM isotype anti-chondroitin sulphate, Sigma)

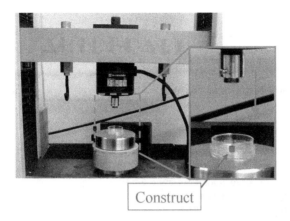

Figure 2. Photograph of compression test for determining the tangent modulus

for 90 min at 37 degree C to primarily label the collagens and the proteoglycan at once. The slices were then washed three times in PBS(–) for 10 min and incubated in PBS(–) containing the three secondary antibodies corresponding to each of the primary antibodies (Alexafluor 350-conjugated anti-mouse IgG1 antibody, A21120; Alexafluor 488-conjugated anti-mouse IgG2a antibody, A11001; Alexafluor 568-conjugated anti-mouse IgM antibody, A21043, Invitrogen) for 60 min at 37 degree C to fluorescently visualize the labeled ECM molecules within the cultured cell-agarose construct. The fluorescently stained specimen was mounted on the coverslip and observed using a confocal laser scanning microscope (CLSM; Eclipse; Nikon Corp., Tokyo, Japan).

2.5. Statistical analysis

Result of the tangent modulus were expressed as the mean ± sample standard deviation (SD). The significance of the difference between each experimental group was assayed using the two-tailed Welch's t-test. The degree of freedom is abbreviated as df. If the results of pairwise comparisons between two groups were $P < 0.01$, $d > 1$ and $(1 - \beta) > 0.8$ simultaneously, we judged the difference in the tangent modulus as significant, where P: level of significance; d: Cohen's d, a kind of the effect size [27]; $(1 - \beta)$: the power of test as calculated usigng the R language.

3. Results

3.1. Effect of VC on mechanical properties of the tissue under free-swelling culture conditions when added to the medium

3.1.1. Influence on mechanical properties

The tangent modulus (E) of the four experiment groups (without VC, AsA(0.64), AsA(2.2), and AsA(3.2)) was determined at 22 days of culture in Fig. 3. The tangent moduli of the high AsA dose groups, AsA(2.2) and AsA(3.2), were significantly higher than those of the without AsA group. Moreover, we measured the temporal growth of the construct's tangent modulus to determine the relationship between development of the ECM and mechanical properties. The tangent moduli of all experiment groups after 1 day of culture were almost identical (results not shown) and were equivalent to the tangent modulus of cell-free agarose gel (12.5 ± 0.9 kPa, $n = 10$). The measured tangent modulus (E) was normalized with the tangent modulus at day 1 (E_0) and plotted against the culture periods in Fig. 4. It was evident that the tangent modulus increased with increasing culture period. These results clearly indicate that the growth rate of the tangent modulus is dependent on the AsA concentration of the culture medium. Thus, the tangent modulus of the high AsA dose groups, AsA(2.2), and AsA(3.2), increased faster than that of the low AsA dose group, AsA(0.64).

3.1.2. Histological observation

Typical immunofluorescence images of elaborated ECMs (type I collagen, type II collagen and chondroitin sulfate) after 22 days of culture were compared as shown in Fig. 5. It is clear that in the low-VC concentration groups (without VC and AsA(0.64)), only a small amount of ECM which is consisting of chondroitin sulfate was present, as indicated in Fig. 5(A) and 5(B). By contrast, abundant ECM, chiefly type II collagen, was observed in the constructs of AsA(2.2) and AsA(3.2), as indicated by the arrow-heads in Fig. 5(C) and 5(D). Then, the broadened collagen network interconnected the chondrocytes to create an extended network which exceeded the field of view in the photograph.

3.2. Influence of mechanical stimulation on the development of the ECM and mechanical properties

3.2.1. Measurement of mechanical properties

To evaluate the influence of mechanical stimulation on the cultured constructs, the tangent moduli of each group after 22 days of culture was measured in Fig. 6. A significant difference was noted between the without VC and AsA(2.2) groups compared with the three free-swelling groups, the latter revealing an increased tangent modulus. When applying cyclic compression to the construct with dosing the culture medium with VC, no clear difference was observed between free-swelling and compression AsA groups. By contrast, both tangent moduli of compression A2P dose group (A2P(6.4)) were higher than those of the free-swelling A2P(6.4) group. Each tangent modulus of control groups with A2P was ranked between free-swelling and compression groups. No differences were evident between the three compression groups ($df = 16$, $P = 0.47$, $d = 0.32$, $(1 - \beta) = 0.030$).

3.2.2. Histological observation

Figs. 7 and 8 show immunofluorescent images at low- and high-magnification and reveal the ECM distribution (type I and II collagen and chondroitin sulfate) after 22 days of culture, with the high-magnification images taken near the center of the construct. It is clear that in the absence of VC in the culture medium, there was only a limited quantity of ECM, as indicated in Fig. 7(A) and 8(A). By contrast, in the free-swelling group (AsA(2.2)), there was fairly uniform distribution of ECM across the construct (Fig. 7(B)) which, at high magnification, revealed a collagen network that interconnected between chondrocytes, as indicated by the arrows in Fig. 8(B). The low-magnification images for each of the A2P dose groups revealed similar distribution in ECM across the constructs (Figs. 7(C)–(E)). In the high-magnification

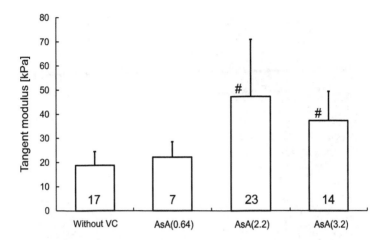

Figure 3. Comparison of tangent modulus after 22 days culture period. Each number in a column is cultured sample number, and each sample represents an individual culture. Error bar means SD. Sharps (#) was shown statistical significant respective as compared to control group and between each groups. #, *: $d > 1$, $P < 0.01$, $(1 - \beta) > 0.8$.

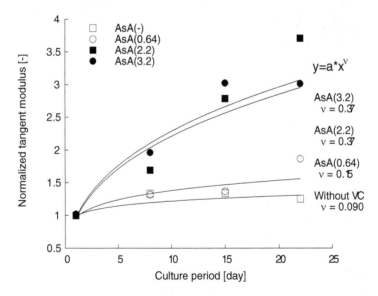

Figure 4. Comparison of normalized tangent modulus with the tangent modulus at day 1 (E_0).

images, type II collagen molecules were appeared to interconnect between chondrocytes, as indicated by arrows in Fig. 8(E). These results demonstrate that the compression group of

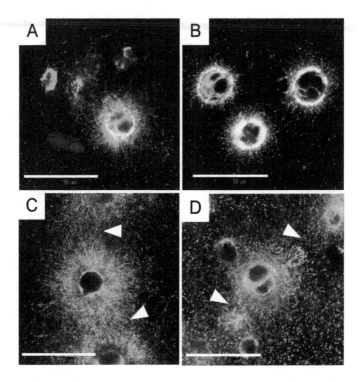

Figure 5. Fluorescence images of ECMs: blue: type I collagen; green: type II collagen; red: chondroitin sulphate. Culture period was 22 days. (A): without VC; (B): AsA(0.64); (C): AsA(2.2); (D): AsA(3.2). Each trichrome stained sample represents an individual culture. Scale bar represents 50 μm.

A2P(6.4) is associated with more of the collagen network than the corresponding free-swelling group. In addition, the distribution of chondroitin sulfate was clearly observed to be restricted to the peripheral regions of the chondrocytes.

4. Discussion

4.1. Effect of VC on mechanical properties of the tissue under free-swelling culture conditions when added to the medium

Collagen synthesis is enhanced by increasing VC concentration [23, 24, 28, 29]. In addition, the immunofluorescent images in Fig. 5 indicate that the synthesized collagen fibrils, as the chief reinforcing fiber in the elaborated tissue, are upregulated by increasing AsA concentration. Moreover, the self-assembled collagen network was spread out spatially in the construct and interconnected among chondrocytes. As shown in Fig. 5, the elaborated ECM is forms into a large network which spreads infinitely. This means that entanglements between opposite sides of a cube in three-dimension exist, and that the entire construct is percolated by interconnecting the ECM network, as shown in Fig. 9. Therefore, the mechanical property of the developed ECM network appeared as that of the whole construct.

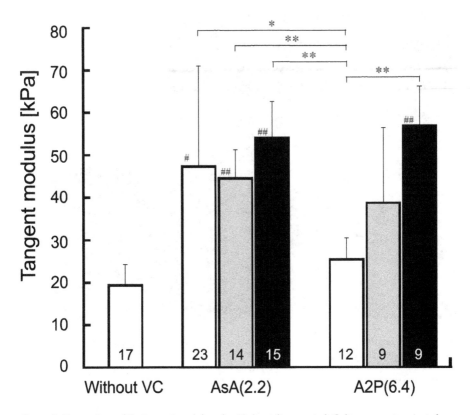

Figure 6. Comparison of the tangent modulus after 22 day culture period. Cultures were terminated following either a free-swelling (white column), control (gray column) or compression (black column) condition. Each number in a column is cultured sample number and each sample represents an individual culture. Error bar means SD. Sharps (# and ##) and asterisks (∗ and ∗∗) were shown statistical significant respective as compared to control group and between each groups in Welch's t-test. #, ∗: $d > 1, P < 0.01, (1 - \beta) > 0.8$; ##, ∗∗: $d > 2, P < 0.01, (1 - \beta) > 0.8$.

The observed mechanical behavior of the cultured constructs, in which the elastic modulus changed non-linearly with the culture period, was similar to the mechanical characteristics of the gel at the vicinity of the sol-gel transition point [30]. In particular, the elasticity of the gel in the vicinity of the sol-gel transition point is exponentially proportional to the molecular concentration of polymer solution [31]. It was widely alleged that the gel had been formed by the very tenuous network irrespective of the mechanism of cross-linking formation process and had been dominated by entropic elasticity in the vicinity of the gelation point Several studied several researches with experiments [31, 32] and numerical simulations [33–35].

On the basis of the above, the chondrocytes synthesize collagen fibrils around themselves, and reinforced polymer concentration (ρ) in the cultured construct is enhanced by increasing collagen density. Then, the elasticity was increased exponentially by culture period (t) as

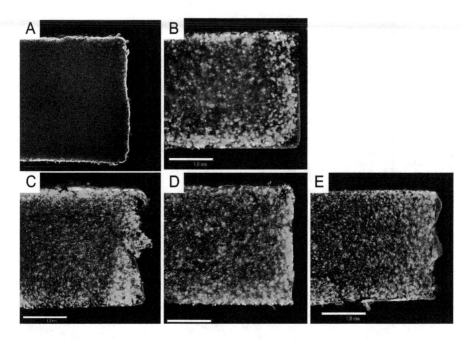

Figure 7. Low-magnified immunofluorescence images of ECMs under either free-swelling (A, B, C), control (D) or compression (E) culture condition: A: without VC; B: AsA(2.2); C–E: A2P(6.4). Blue: type I collagen; green: type II collagen; red: chondroitin sulphate. Culture period was 22 days. Each trichrome stained samples represents an individual culture. Scale bar represents 1 mm.

follow,

$$E \propto |\rho - \rho_c|^\lambda \propto t^\nu, \tag{1}$$

where ρ_c is polymer density at gelation point. We considered that the elastic modulus of the regenerated cartilage is expected to exponentially proportionate with the culture period. Therefore, we hypothesised that the tangent modulus of cultured construct increases exponentially with the culture period because the ECM content of the cultured constructs improved with the culture period by the following relationship,

$$\frac{E}{E_0} \propto t^\nu, \tag{2}$$

where E_0 is tangent modulus at day 1 of the culture period and exponent ν is the growth rate of the tangent modulus. To investigate the relationship between AsA concentration and the growth rate of the tangent modulus, we applied equation (2) to the tangent moduli of all experimental groups in Fig. 4 and described these exponents (ν) into Fig. 10. As shown in Fig. 10, the exponent (ν) was linearly proportional to the AsA concentration with 3.2 pmol/10^9 cells as follows,

$$\nu \propto (\text{AsA concentration}). \tag{3}$$

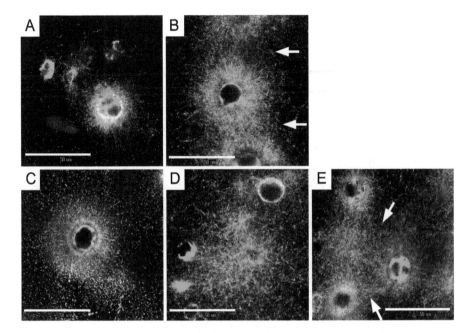

Figure 8. High-magnified immunofluorescence images of ECMs in the vicinity of the tissue model under either free swelling (A, B, C), control (D) or compression (E) culture condition: A: without VC; B: AsA(2.2); C–E: A2P(6.4). Blue: type I collagen; green: type II collagen; red: chondroitin sulphate. Culture period was 22 days. Each trichrome stained samples represents an individual culture. Scale bar represents 50 μm. Arrows indicate to interconnect among chondrocytes by type II collagen net work.

To establish a suitable design method for regenerating tissue-engineered cartilage, we studied the effect of two types of chemical and physical stimulation on the mechanical properties of the chondrocyte-agarose construct as a regenerated cartilage model. One is to add a high quantity of VC to the culture medium to effectively enhance collagen synthesis. The other is to apply a compressive strain to the construct using a purpose-built bioreactor. Our results showed that the compressive strain led to an increase in the tangent modulus because it was clear that the collagen network had increased its density and had interconnected chondrocytes.

4.2. Influence of mechanical stimulation on the development of the ECM and mechanical properties

To establish a suitable design method for regenerating tissue-engineered cartilage, we studied the effect of two types of chemical and physical stimulations on the mechanical property of the chondrocyte-agarose construct, as a regenerated-cartilage tissue model. One is to add high VC concentration into culture medium for effectively enhancing the collagen synthesis. The other is to apply a compressive strain to the construct using a purpose-built bioreactor. Our results showed that compressive strain had leaded to increase tangent modulus because it

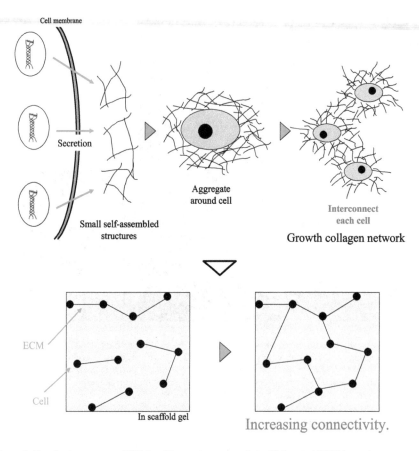

Figure 9. Developing process of ECM and increasing connectivity. Elaborated ECM forms large network which is spreaded infinitly. This means a entanglement between opposite sides of a cube in three dimensions, and that entire mechanical property was percolated by of interconnected ECM network.

had been clear that collagen network had become dense and had been interconnected among chondrocytes. We already described in last section (4.1) that it had definitely important that chondrocytes had been linked by ECM for development of the elasticity of the construct.

It is well-known that A2P has no physiological activity but can produce the same as that of effect of VC activity after dephosphorylation by an alkaline phosphatase (ALP) [36]. Dephosphorylated A2P, this is AsA, penetrates into chondrocyte through a VC transporter, chiefly the sodium-dependent VC transporter 2 (SVCT2), and supports the collagen synthesis as a cofactor in the rough endoplasmic reticulum [37]. Under free-swelling culture conditions, we observed few collagen molecules distributed in the construct and found that the tangent modulus of the A2P group was lower than that of the AsA(2.2) group. We should also consider reaction rates, one of A2P dephosphorylation by ALP, the other of AsA transport into the

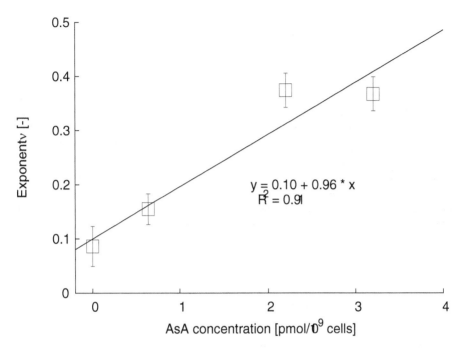

Figure 10. Relationship of AsA concentration and exponent. Line shows a linear regression in AsA concentration interval of 0 to 3.2 pmol/10^9 cells. R is the coefficient of determination.

cytosol by SVCT2. The Michaelis constants of both ALP and SVCT2 of bovine chondrocyte is 1–10 and $62 \pm 25\mu$M [38, 39]. Since the affinity of the substrate for ALP is slightly higher than for SVCT2, AsA concentration around chondrocyte in A2P group was lower unlike cultivation of high AsA concentration, the also rate of collagen synthesis was relatively decreased by comparison with AsA dose group. Then, the tangent modulus of the free-swelling groups with the A2P was suppressed with restraining cytotoxicity of AsA.

When applying compressive strain to the construct, the tangent modulus of A2P(6.4) was higher than that of the respective free-swelling groups. This is because the collagen fibers interconnected chondrocytes and because the collagen network of these groups was expanded. We think this mechanical stimulation enhanced the diffusion of both A2P and nutrients, homogenizing A2P in the construct. In addition, the stimulation probably excited a mechanosensor on the cell surface, activating cell-signaling pathways, namely the mechanotransduction pathway. It is well-known that mechanical stimulation causes chondrocytes to several biological responses in cartilage remodeling strategies which are based on the implantation of a cultured tissue [40]; for example, activation of the mitogen-activated protein kinase (MAPK) pathway, a kind of cell signaling pathway [41, 42], increase in GAG biosynthesis [43] and regulation of inflammatory species synthesis [44]. These chondrocyte responses were followed presently after loading mechanical stimulation and sustained the activity of MAPK pathway for 5–60 min [41, 42]. BY synthesizing ECM with mechanical stimulation, chondrocytes built up the collagen network and adapted to changes in the deformation around the cells. Therefore, the tangent modulus of the compression group with A2P was higher than that of the free-swelling group.

Homogenously developing ECM under a free-swelling culture condition for expansive regeneration of cartilage is a complex process. Our results provide evidence that it is necessary to subject the construct to mechanical strain because nutrients should be supplied uniformly to the construct. We also think that a diffusion of nutrients exists under free-swelling culture conditions. The diffusion coefficient of water in agarose gel is about $10 \times 10^{-10} m^2 s$ [45, 46]. This means that a water molecule diffuses 10 mm per a day in the gel when applying the Fick's law. Pluen *et al.* reported that the diffusion coefficients of the proteins, lactalbumin and ovalbumin, in 0.1 M PBS(–) solution into agarose gel was 0.8–1 $\times 10^{-10} m^2 s$ [47]. This means that the nutrient particles of diffuse 3 mm per a day in the gel. Moreover, if the ECM of the cultured construct is more densely synthesized, it is hard to diffuse nutrients in the tissue; and consequently, the diffusion coefficient will be decreased. Thus, if a medical doctor would like to implant a large quantity of regenerated cartilage to treat the cartilage defect, the construct should be subjected to mechanical stimulation to homogenously develop the ECM network.

5. Summary

This chapter demonstrates the relationship between the development of the ECM in the construct and the mechanical properties of the construct for establishing a suitable design method to reconstruct the regenerated cartilage by using a chondrocyte-agarose construct as a tissue model. First, we revealed the influence of VC concentration in the culture medium on the mechanical properties of the regenerated cartilage model. The present findings suggest that the mechanical characteristic of the construct depend clearly on the AsA concentration in the culture medium. The tangent modulus of the cultured construct was exponentially increased according to cultivation duration. The growth rate of the tangent modulus was accelerated to upregulate ECM secretion in the high AsA concentration. This study strongly suggests, therefore, that the mechanical properties of the regenerated cartilage depend on the interconnections created by the ECM between cells which are mainly those of the three-dimensional collagen network. Second, we investigated the influence of mechanical properties on the development of the ECM network as well as the regulation of the chondrocyte-agarose constructs using two types of VCs, AsA and A2P. Neither the collagen network nor the tangent modulus of the A2P dose groups was improved compared with the AsA group under the free-swelling culture condition. Moreover, it is clear that the free-swelling culture condition suppresses the development of the ECM of the inner tissue compared with the ECM of outer tissue. W hen applying compressive strain to the construct, the tangent modulus of the A2P dose group was increased because the ECM networks of the inner tissue had been upregulated and had interconnected chondrocytes. We can additionally consider that mechanical stimulation enhanced the diffusion of nutrients and improved the synthesis of the ECM via the mechanotransduction pathway. Moreover, we revealed that it is necessary to apply mechanical stimulation to a large engineered tissue when treating articular cartilage defects.

Acknowledgments

This study was financially supported by the Grant-in-Aid for Scientific Research of Japan Society for the Promotion of Science (JSPS, No. 20360078).

Author details

Seiji Omata
Department of Biomedical Engineering, Graduate School of Biomedical Engineering, Tohoku University, Japan

Yoshinori Sawae
Department of Mechanical Engineering, Faculty of Engineering, Kyushu University, Japan

Teruo Murakami
Advanced Biomaterials Division, Research Center for Advanced Biomechanics, Kyushu University, Japan

6. References

[1] Waters, T. & Bentley, G. (2007). Articular cartilage. *Basic Orthopaedic Sciences: The Stanmore Guide*, E. Arnold, (Ed.), pages 85–94. Hodder Arnold, ISBN-13: 978-0340885024.

[2] Kheir, E. & D. Shaw, D. (2009). Hyaline articular cartilage. *Orthopaedics and Trauma*, Vol. 23, No. 6, pp. 450–455, ISSN 1877-1327.

[3] Treppo, S.; Koepp, H.; Quan, E. C.; Cole, A. A.; Kuettner, K. E.; Grodzinsky, A. J. (2000). Comparison of biomechanical and biochemical properties of cartilage from human knee and ankle pairs. *Journal of Orthopaedic Research*, Vo. 18, No. 5, pp. 739–748, ISSN 1554-527X.

[4] Schulz, R. M. & Bader, A. (2007). Cartilage tissue engineering and bioreactor systems for the cultivation and stimulation of chondrocytes. *European Biophysics Journal*, Vol. 35, No. 4–5, pp. 539–568, ISSN 1432-1017.

[5] McNary, S. M.; Athanasiou, K. A.; Reddi, A. H. (2012). Engineering lubrication in articular cartilage. *Tissue Engineering: Part B*, Vol. 18, No. 2, pp. 88–100, ISSN 2152-4955.

[6] Mow, V.C. & Guo, X.E. (2002). Mechano-electrochemical properties of articular cartilage: Their inhomogeneities and anisotropies. *Annual Review of Biomedical Engineering*, Vol. 4, pp. 175–209, ISSN 1523-9829.

[7] Schiavinato, A. & Whiteside, R. A. (2011). Effective lubrication of articular cartilage by an amphiphilic hyaluronic acid derivative. *Clinical Biomechanics*, In Press, ISSN 0268-0033.

[8] Getgood, A.; Brooks, R.; Fortier, L.; Rushton, N. (2009). Articular cartilage tissue engineering *The Journal of Bone and Joint Surgery (British)*, Vol. 91, No. 5, pp. 565–576, ISSN2044-5377.

[9] Frenkel, S. R. & Di Cesare, P. E. (2004). Scaffolds for articular cartilage repair. *Annals of Biomedical Engineering*, Vol. 32, No. 1, pp. 26–34, ISSN 1573-9686.

[10] Farr, J.; Cole, B.; Dhawan, A.; Kercher, J.; Sherman, S. (2011). Clinical cartilage restoration: Evolution and overview. *Clinical Orthopaedics and Related Research*, Vol. 469, No. 10, pp. 2696–2705, ISSN 1528-1132.

[11] Revell, C. M. & Athanasiou, K. A. (2009). Success rates and immunologic responses of autogenic, allogenic, and xenogenic treatments to repair articular cartilage defects. *Tissue Engineering: Part B*, Vol. 15, No. 1, pp. 1–15, ISSN 2152-4955.

[12] Gomoll, A. H.; Filardo, G.; de Girolamo, L.; Esprequeira-Mendes, J.; Marcacci, M.; Rodkey, W. G.; Steadman, R. J.; Zaffagnini, S.; Kon, E. (2012). Surgical treatment for

early osteoarthritis. part I: cartilage repair procedures. *Knee Surgery, Sports Traumatology, Arthroscopy*, Vol. 20, No. 3, pp. 450–466, ISSN 1433-7347.

[13] Gomoll, A. H.; Filardo, G.; Almqvist, F. K.; Bugbee, W. D.; Jelic, M.; Monllau, J. C.; Puddu, G.; Rodkey, W. G.; Verdonk, P.; Verdonk, R.; Zaffagnini, S.; Marcacci, M. (2012). Surgical treatment for early osteoarthritis. part II: allografts and concurrent procedures. *Knee Surgery, Sports Traumatology, Arthroscopy*, Vol. 20, No. 3, pp. 468–486, ISSN 1433-7347.

[14] Omata, S.; Sawae, Y.; Murakami, T.. (2011). Influence of ascorbic acid (AsA) concentration in culture medium on mechanical property of regenerated cartilage. *Journal of Environment and Engineering*, Vol. 6, No. 2, pp. 416–425 ISSN 1880-988X.

[15] Mandl, J.; Szarka, A.; Bánhegyi, G. (2009). Vitamin C: Update on physiology and pharmacology. *British Journal of Pharmacology*, Vol. 157, No. 7, pp. 1097–1110, ISSN 1476-5381.

[16] Shoulders, M. D. & Raines, R. T. (2009). Collagen structure and stability. *Annual Review of Biochemistry*, Vol. 78, pp. 929–958, ISSN 1545-4509.

[17] Gelse, K.; Pöschl, E.; Aigner, T. (2003). Collagens — structure, function, and biosynthesis. *Advanced Drug Delivery Reviews*, Vol. 12, No. 28, pp. 1531–1546, ISSN 0169-409X.

[18] Nagata, K. (1996). Hsp47: a collagen-specific molecular chaperone. *Trends in Biochemical Sciences*, Vol. 21, No. 1, pp. 23–26, ISSN 0968-0004.

[19] Vunjak-Novakovic, G.; Martin, I.; Obradovic, B.; Treppo, S.; Grodzinsky, A. J.; Langer, R.; Freed, L. E. (1999). Bioreactor cultivation conditions modulate the composition and mechanical properties of tissue-engineered cartilage. *Journal of Orthopaedic Research*, Vol. 17, No. 1, pp. 130–138, ISSN 1554-527X.

[20] Mauck, R. L.; Seyhan, S. L.; Ateshian, G. A.; Hung, C. T. (2003). Influence of seeding density and dynamic deformational loading on the developing structure/function relationships of chondrocyte-seeded agarose hydrogels. *Annals of Biomedical Engineering*, Vol. 30, No. 8, pp. 1046–1056, ISSN 1573-9686.

[21] Kelly, D. J.; Crawford, A.; Dickinson, S. C.; Sims, T. J.; Mundy, J.; Hollander, A. P.; Prendergast, P. J.; Hatton, P. V. (2007). Biochemical markers of the mechanical quality of engineered hyaline cartilage. *Journal of Materials Science: Materials in Medicine*, Vol. 18, No. 2, pp. 273–281, ISSN 1573-4838.

[22] Ronziére, M.-C.; Farjanel, J.; Freyria, A.-M.; Hartmann, D. J.; D. Herbage, D. (1997). Analysis of types I, II, III, IX and XI collagens synthesized by fetal bovine chondrocytes in high-density culture. *Osteoarthritis and Cartilage*, Vol. 5, No. 3, pp. 205–214, ISSN 1063-4584.

[23] Clark, A. G.; Rohrbaugh, A. L.; Otterness, I.; Kraus, V. B. (2002). The effects of ascorbic acid on cartilage metabolism in guinea pig articular cartilage explants. *Matrix Biology*, Vol. 21, No. 2, pp. 175–184, ISSN 0945-053X.

[24] Ronziére, M.-C.; Roche, S.; Gouttenoire, J.; Démarteau, O.; Herbage, D.; Freyria, A.-M. (2003). Ascorbate modulation of bovine chondrocyte growth, matrix protein gene expression and synthesis in three-dimensional collagen sponges. *Biomaterials*, Vol. 24, No. 5, pp. 851–861, ISSN 0142-9612.

[25] Bean, A. C.; Almarza, A. J.; Athanasiou, K. A. (2006). Effects of ascorbic acid concentration on the tissue engineering of the temporomandibular joint disc. *Proceedings of the Institution of Mechanical Engineers, Part H: Journal of Engineering in Medicine*, Vol. 200, No. 3, pp. 439–447, ISSN 2041-3033.

[26] Lee, D. A. & Bader, D. L. (1997). Compressive strains at physiological frequencies influence the metabolism of chondrocytes seeded in agarose. *Journal of Orthopaedic Research*, 15(2):181–188, ISSN 1554-527X.

[27] Cohen, J. (1988). *Statistical power analysis for the behavioral sciences — 2nd ed.* Lawrence Erlbaum Assoc Inc., ISBN-13: 978-0805802832.

[28] Tajima, S. & Pinnell, S. R. (1982). Regulation of collagen synthesis by ascorbic acid. Ascorbic acid increases type I procollagen mRNA. *Biochemical and Biophysical Research Communications*, Vol. 106, No. 2, pp. 632–637, ISSN 0006-291X.

[29] Lyons, B. L. & Schwarz, R. I. (1984). Ascorbate stimulation of pat cells causes an increase in transcription rates and a decrease in degradation rates of procollagen mRNA. *Nucleic Acids Research*, Vol. 12, No. 5, pp. 2569–2579, ISSN; 1362-4962.

[30] Axelos, M. A. V. & Kolb, M. (1990). Crosslinked biopolymers: Experimental evidence for scalar percolation theory. *Physical Review Letters*, Vol. 64, No. 12, pp. 1457–1460, ISSN 1079-7114

[31] Tokita, M. & Hikichi, K. (1987). Mechanical studies of sol-gel transition: Universal behavior of elastic modulus. *Physical Review A*, Vol. 35, No. 10, pp. 4329–4335, ISSN 1094-1622

[32] Fujii, T.; Yano, T.; Kumagai, H.; Miyawaki, O. (2000). Scaling analysis of the concentration dependence on elasticity of agarose gel. *Bioscience, Biotechnology, and Biochemistry*, Vol. 64, No. 8, pp. 1618–1622, ISSN: 1347-6947.

[33] Farago, O. & Kantor, Y. (2002). Entropic elasticity at the sol-gel transition. *Europhysics Letters*, Vol. 57, No. 3, pp. 458–463, ISSN 1286-4854.

[34] Plischke, M. (2006). Critical behavior of entropic shear rigidity. *Physical Review E*, Vol. 73, No. 6, 61406, 2006, ISSN 1550-2376.

[35] Surve, M.; Pryamitsyn, V.; Ganesan, V. (2006). Universality in structure and elasticity of polymer-nanoparticle gels. *Physical Review Letters*, Vol. 96, No. 17, 177805, ISSN 1079-7114.

[36] Takamizawa, S.; Maehata, Y.; Imai, K.; Senoo, H.; Sato, S.; Hata, R. (2004). Effects of ascorbic acid and ascorbic acid 2-phosphate, a long-acting vitamin C derivative, on the proliferation and differentiation of human osteoblast-like cells. *Cell Biology International*, Vol. 27, No. 4, pp. 255–262, ISSN 1065-6995.

[37] Savini, I.; Rossi, A.; Pierro, C.; Avigliano, L.; Catani, M. V. (2008). SVCT1 and SVCT2: key proteins for vitamin C uptake. *Amino Acids*, Vol. 34, No. 3, pp. 347–355, ISSN 0939-4451.

[38] McNulty, A. L.; Vail, T. P.; Kraus, V. B. (2005). Chondrocytenext term transport and concentration of ascorbic acid is mediated by SVCT2. *Biochimica et Biophysica Acta (BBA) — Biomembranes*, Vol. 1712, No. 2, pp. 212–221, ISSN 0005-2736.

[39] Fortuna, R.; Anderson, H. C.; Carty, R. P.; Sajdera, S. W. (1979). Enzymatic characterization of the chondrocytic alkaline phosphatase isolated from bovine fetal epiphyseal cartilage. *Biochimica et Biophysica Acta (BBA) — Enzymology*, Vol. 570, No. 2, pp. 291–302, ISSN 0005-2744.

[40] Ramage, L.: Nuki, G.; Salter, D. M. (2009). Signalling cascades in mechanotransduction: cell-matrix interactions and mechanical loading. *Scandinavian Journal of Medicine & Science in Sports*, Vol. 19, No. 4, pp. 457–469, ISSN 1600-0838.

[41] Fitzgerald, J. B.; Jin, M.; Dean, D.; Wood, D. J.; Zheng, M. H.; Grodzinsky, A. J. (2004). Mechanical compression of cartilage explants induces multiple time-dependent gene

expression patterns and involves intracellular calcium and cyclic amp. *The Journal of Biological Chemistry*, Vol. 279, No. 19, pp. 19502–19511, ISSN 1083-351X.

[42] Fitzgerald, J. B.; Jin, M.; Chai, D. H.; Siparsky, P.; Fanning, P.; Grodzinsky, A. J. (2008). Shear- and compression-induced chondrocyte transcription requires MAPK activation in cartilage explants. *The Journal of Biological Chemistry*, Vol. 283, No. 11, pp. 6735–6743, ISSN 1083-351X.

[43] Kisiday, J. D.; Lee, J.H.; Siparsky, P. N.; Frisbie, D. D.; Flannery, C. R.; Sandy, J. D.; Grodzinsky, A. J. (2009). Catabolic responses of chondrocyte-seeded peptide hydrogel to dynamic compression. *Annals of Biomedical Engineering*, Vol, 37, No. 7, pp. 1368–1375, ISSN 1573-9686.

[44] Akanji, O. O.; Sakthithasan, P.; Salter, D. M.; Chowdhury, T. T. (2010). Dynamic compression alters NFκB activation and IκB-α expression in IL-1β-stimulated chondrocyte/agarose constructs. *Inflammation Research*, Vol. 59, No. 1, pp. 41–52, ISSN 1420-908X.

[45] Davies, E.; Huang, Y.; Harper, J. B.; Hook, J. M.; Thomas, D. S.; Burgar, I. M.; Lillford, P. J. (2010). Dynamics of water in agar gels studied using low and high resolution 1h nmr spectroscopy. *International Journal of Food Science & Technology*, Vol. 45, No. 12, pp. 2502–2507, ISSN 1365-2621.

[46] Gustafsson, N. O.; Westrin, B.; Axelsson, A.; Zacchi, G. (1993). Measurement of diffusion coefficients in gels using holographic laser interferometry. *Biotechnology Progress*, Vol. 9, No. 4, pp. 436–441, ISSN 1520-6033.

[47] Pluen, A.; Netti, P. A.; Jain, R. K.; Berk, D. A. (1999). Diffusion of macromolecules in agarose gels: comparison of linear and globular configurations. *Biophysical Journal*, Vol. 77, No. 1, pp. 542–552, ISSN 1542-0086.

Chapter 8

Placental Structure and Biological Aspects of Fetal Membranes Cultured *in vitro*

João Bosco Barreto Filho and Maira Souza Oliveira

Additional information is available at the end of the chapter

1. Introduction

The placenta is a transient organ responsible for maternal fetus nutrient and oxygen exchange. It is important to the modulation of the maternal immune response to antigens of paternal and fetal origin and a source of a great variety of hormones that ensure the maintenance of gestation. Pathological conditions of pregnancy, fetal growth and delivery, abnormal placentation and surgical problems always attracted the medical sciences attention. The diversity of placental types among eutherian mammals has often raised important questions about phylogeny. *In vitro* culture of placental cells is an approach that allows the investigation of hormonal production and metabolic process as well as pathological disorders of common occurrence in human beings and animals. This chapter aims to show the most important features on culturing placenta derived cells and cell populations with properties of progenitor/stem cells: identification of the structures, isolation of the cells, maintenance of the cell culture, and applicability in research trends.

2. The placenta

Retention and development of the fertilized egg within the maternal body (histotrophic and hemotrophic viviparity) are observed in most mammals, reptiles and a few lower organisms. The growing embryo is nourished by the mother, usually through a placenta or similar structure. Vertebrate viviparity with the development of a placenta in the uterus evolved 150-100 million years ago and is responsible for the adaptability and dispersion of eutherian mammalians all over the diverse habitats.

The Mammalian class is divided into two subclasses, Prototheria and Theria, which includes the two major groups of viviparous mammals, the marsupials and placentals.

Four superorders of eutherian mammals were identified by techniques of molecular phylogenetics. The two oldest superorders Afrotheria (elephants and others) and Xenarthra (armadillos, anteaters and sloths) have either endothelial or hemochorial placentas. Members of the superorder Euarchontoglires exhibit hemochorial placentas (Glires – rodents and lagomorphs) and endotheliochorial placentation (Euarchonta – tree shrews). Epitheliochorial placentas are observed in the last superorder to arise, the Laurasiatheria (horses, ruminants and various other species). It seems that the likely path of evolution in Afrotheria was from endotheliochorial to hemochorial placentation and the same condition apparently occurs with the Xenarthra and the bats. It is possible that endotheliochorial placentation was the more primitive and that epitheliochorial placenta evolved twice, once in the Laurasiatheria and once in Euarchontoglires. Mammalian phylogeny has been extensively studied (1–4).

The placenta should be considered as an apposition of the fetal membranes to the uterine tissue for physiological exchange, and in mammals the definitive placentation is of the chorio-allantoic type, in which mesodermal blood vessels, from the allantoids external layer, form a vascular bridge between the embryo and the chorion, allowing a broad vascularization, and increasing efficiency in the mother-fetus exchange (5). In this situation, the maternal-fetal interface is enlarged, because chorionic villous fills the uterine crypts. In the chorion-vitelline placenta, there is a fusion between the yolk sac and its blood vessels network with the chorion. Here the allantois is never in contact with the chorion and it is observed in some non-mammalian vertebrates. In mammalian species, however, it occurs temporarily, while the allantois is developing from the intestine, to constitute the chorio-allantoic placenta.

Despite of all the work to develop systems of classification and to use placental structure to aid in the investigation of phylogenetic relationships among species of different orders, there is a lack of morphological and electron microscopic studies of the placenta in a great number of organisms, limiting the efforts to generalizations. The classical systems of classification of the placenta, based on structure, are of limited significance. Considering that the trophoblast could be apposed to uterine ephithelium, to the endothelium of maternal vessels or directly to maternal blood, the placenta is classified as epitheliochorial, endotheliochorial or hemochorial, respectively. However, electron microscopy studies have been shown that a continuous fetal endothelium and chorion always persist, although very thin in some tissue extensions, reducing the diffusion distance between fetal and maternal circulation, even in epitheliochorial placentas.

It is now well recognized that the number of layers between fetal and maternal circulation do not have any relationship to the placenta's ability to supply oxygen to the fetus. Interhaemal area is reduced in many ways in the different types of placentation. In the ruminant's placenta (epitheliochorial and synepitheliochorial) the indentation of trophoblast and uterine epithelium by blood vessels decreases the interhaemal distance; in the endotheliochorial placentae this reduction is achieved by the indentation of trophoblast by fetal capillaries within the labyrinth. In the haemochorial placentation several features occur to reduce the thickness of the trophoblast, like alternation of thick and thin regions in the rabbit placenta (See (6) for detailed information).

The placenta developed as an organ to nutrient and oxygen exchange between the mother and fetus and also to avoid immunological rejection by the maternal immune system. Cells from the mother and the fetus intermingle throughout gestation and the extension of trophoblast invasion apparently is related to the modulation of maternal immune response against the allogenic conceptus (7). In the epithelialchorion placentation there are minimal changes in the uterine mucosa during gestation, whereas in the haemochorial placentation the endometrium is differentiated into decidua. These changes allow the establishment and maintenance of pregnancy and fetal survival with great diversity among different species.

2.1. Organogenesis

At the beginning of the mammalian development, the conceptus differentiates into an inner cell mass and an outer layer of cells, the trophoblast, which solely contributes to extra-embryonic membranes formation (4,8). The tissue on the maternal component of the placenta usually is of epithelial or connective tissue origin of the ovary, oviduct or uterus. The fetal component is a derivate of ectodermal epithelium, the trophoblast or trophectoderm.

Diverse populations of trophoblasts, all derived from the embryonic trophectoderm, have morphological, functional and molecular diversity within and across species. The extra-embryonic membranes of the vertebrate conceptus form from the three germ layers of the embryo - the endoderm, mesoderm and ectoderm. These germ layers form the yolk sac, initially a single layer of ectoderm that is progressively transformed in a bilaminar and trilaminar structure, after the migration of a layer of endoderm internal to the ectoderm, and a growth of mesoderm between these two first layers. In mammals, the growth of mesoderm transforms only the superior part of the bilaminar yolk sac into the trilaminar vascularized structure, which is the basis for development of the chorion, amnion and allantois. Fusion of mesoderm and endoderm vascularizes promptly, whereas mesoderm and trophectoderm rarely form bloof vessels. The vascularized yolk sac fused with the ectoderm form the fetal component of the choriovitelline placenta, and the vascularized allantois plus chorion forms the fetal part of the chorioallantoic placenta.

The trophectoderm and the inner cell mass become separate during early gestation and non interconvertible cell lineages. In mouse, for instance, Cdx2 is expressed predominantly in the trophoblast and Oct3/4 only in the inner cell mass (6). In the human placenta (9), the cell types that constitute placental villi are different populations of trophoblasts, the syncytiotrophoblasts and cytotrophoblasts that cover the surface of villi and bathe in maternal blood within intervillous space. Trophoblasts are formed during the first stage of pregnancy and are the first cells to differentiate from the fertilized egg. They have two cell populations, the undifferentiated cytotrophoblasts and fully differentiated syncytiotrophoblasts. The syncytiotrophoblasts are a continuous, specialized layer of epithelial cells.

Mesenchymal cells, Hofbauer cells (mesenchymal derived macrophages) and fibroblasts are found between fetal vessels and trophoblasts in human species. The last group of cells is fetal vascular cells (vascular smooth muscle cells, endothelial and perivascular cells (pericytes). Hofbauer cells are the macrophages in the placenta villous stroma. These cells are of mesenchymal origin and expand during the first and second trimesters in placental villous tissues mesenchymal stem cells are differentiated by vasculogenesis and angiogenesis, transform into hemangioblastic cell cords, which are believed to be the precursors of capillary endothelial cells and hematopoietic stem cells. Mesenchymal stem cells can also differentiate into perivascular cells, which are considered predecessors of capillary endothelial cells. Pericytes (perivascular cells with dendritic processes) are found surrounding capillary endothelium and venules. They support endothelial cells and are important for maintaining vessel stability and microvascular integrity.

In mouse, the trophectoderm differentiates in two pathways: rapid proliferation of polar trophoblastic cells originate the extraembrionic ectoderm and the ectoplacental cone; and the mural trophoblastic cells that originate the primary giant trophoblastic cells by endoreplicating their DNA but not dividing, like in other rodent giant cells. These cells, however, are not analogous to that observed in the human placenta, despite bearing the same name.

Chorionic binucleate cells are observed in the ruminant and horse placentas (6). In the mare, these are migratory transient cells and never fuse with maternal cells. In the cow and ewe, nevertheless, this population of fetal cells migrates through the chorionic tight junction to fuse with uterine epithelial cells throughout gestation. These cells are derived from uninucleate trophectoderm cells by consecutive nuclear divisions, the second without subsequent cytokinesis. The chorionic binucleate cells comprise 15 to 20 per cent of the trophectoderm epithelial cells and produce a great variety of hormones. This is a unique feature of the ruminant placenta.

2.2. Structure and classification

Placenta has been described in the early studies (5) as more variable in structure than any other mammalian organ, and consequently it is classified in different ways, but other classification (10), which distinguishes the layers between fetal trophoblast and maternal endometrial surface, is considered the most useful, despite its limitations regarding biological and evolutionary aspects, as it was mentioned before.

Morphologically, the placenta is classified according to the configuration of the maternal-fetal interface. The area of fetomaternal exchange is increased by the formation of placental folds (villi) which is characteristic among families. Four types of placenta are described: in the diffuse placenta (horse and pig) the allantochorion is involved in the formation of the organ and it is almost entirely in contact with the endometrium; the cotyledonary placenta, observed in ruminants, shows multiple and discrete areas of attachment to the endometrium. The fetal portion is called cotyledon and the maternal contact sites are the caruncles. The hemotrophic structures formed by this fusion are called placentomes.

The placenta seen in carnivores like dogs and cats is called zonary, and the chorionic villi occupy the equatorial region of the chorionic sac, where they attach to the endometrium;

finally, a discoid placenta is present in rodents and primates, in which an area of the chorion (discoid in shape) adhere to the endometrial stroma.

Regarding to its internal structure, the placenta is classified in villous and labyrinthine; in the former, chorionic villi are present and they penetrate into uterine crypts; fetal blood is transported in vessels through these structures. In the labyrinthine placenta, maternal blood circulates through channels within the fetal syncytiotrophoblast.

Histologically, the placenta is classified according to the layers between fetal and maternal blood, examined by light microscopy. The fetal components are endothelium, mesenchyma and trophoblast; maternal tissues are uterine epithelium, connective tissue and endothelium. When the uterine epithelium is in contact with the chorion (pig and horse) the placenta is called epitheliochorial; in ruminants, the uterine epithelium is removed and the maternal connective tissue is in contact with the chorion (Syndesmochorial placenta); uterine epithelium and connective tissue could be absent, and so the maternal endothelial basement membrane is in contact with the chorion, like in the carnivores placenta, and finally, in the haemochorial placenta, all the maternal tissue layers are removed, the chorion is in direct contact with the maternal circulation and such situation is observed in anthropoids and rodents.

Some other criteria of classification have been proposed, like capillary position, regional specialization, tissue lost at parturition and accessory placental structures. For the classical criteria see (11) and for updates, see (6).

2.3. Physiology

The placenta is related to the production and metabolism of gonadotropins, steroids and prostaglandins hormones that are responsible for the pregnancy establishment and maintenance, placental maturation and parturition. Throughout gestation uterine environment shows great plasticity, which is necessary for fetal growth, being mediated by hormonal production of the placenta. This organ produces a great variety of hormones, including steroid hormones, peptides and insulin-like growth factors. In the human species, these factors are related to proliferation, invasion and differentiation of the trophoblast. Progesterone (P4) secretion in mammals is necessary to prepare the uterine environment to receive the developing conceptus. Embryonic signals, in the absence of P4, are inefficient to rescue the corpus luteum from the uterus luteolytic mechanisms, like in the ungulates, or to extend the luteal lifespan, as it occurs in the human species. In the ewe, hypophysectomy before day 50 of gestation causes corpus luteum regression and abortion. After this, the placenta produces P4 in sufficient concentrations to maintain pregnancy, even in the absence of ovaries and hypophysis.

In ruminants, binucleate cells (BNC) are present in the trophoblast since the pre-implantation period until parturition and these cells are found in the endometrium by day 22 of pregnancy. BNC produce placental lactogen, P4, and prostaglandins E-2 and I-2. In the mare these cells produce the equine chorionic gonadotropin during their migration by the uterine connective tissue. In the cow steroid production are altered throughout gestation. Two to three days before parturition, P4 levels decrease, and estradiol, in contrast, increase, and high levels are observed in the last week of gestation.

BNC produce prostaglandins and it is possible that they are related to the maternal recognition of pregnancy in some species, like swine. It has been suggested that estrogens and prostaglandins have a physiological role in normal expulsion of the placenta in cattle.

In the ruminant placenta, the trophoblast produces interferons that are responsible by the maternal recognition of pregnancy. Interferon tau has been recognized, in ungulates, as an inhibitor of luteolysis. Atypical interferons are produce by the swine blastocyst and its relationship with the maternal recognition of pregnancy is less clear. Type I interferons are produced by the human placenta and it is accepted that they are hormones related to the establishment of gestation in this species. For details see (12,13).

3. *In vitro* culture of placental tissues

3.1. Primary culture

Basically the source of tissues to primary culture is human term placentas or animal placentas obtained in abattoirs. Cells can be obtained by mechanical disaggregation of placental tissues (amniochorion and allantochorion membranes, cotyledons, caruncles and the endometrium itself). Usually tissues are rinsed in Hank's balanced salt solution (HBSS) or in phosphate-buffered saline (PBS) enriched with antibiotics (gentamicin, penicillin, streptomycin), to avoid bacterial contamination, and they are progressively dispersed with surgical scissors in at least six baths. After rinsed, tissues are cultured in suspension in Minimum Essential Medium (EMEM) with Earle's salts. Disaggregate tissues contain different cell types and it may be necessary to go through a separation process. Amniocytes may be aspirated directly from placental fetal fluid and cultured in specialized medium such as Amniomax. These cells have adherent properties and form small aggregates in culture plastic flasks, being ideal for cariotypying. Basic culture conditions are 37º C and 5% CO_2 atmosphere. Briefly, a protocol of cell separation is described. First, remove fat and dead tissues. Wash it with PBS twice. Slice placenta in trophoblast regions and transfer to a sterilized conical tube with cold PBS (10%) on ice. Release trophoblasts from chorionic villi by trypsin digestion. Plate the cells. Deplete macrophages by adherence to plastic. Freshly isolated cells are predominantly trophoblast villous that can be identified by verifying the expression of epidermal growth factor receptor (EGFR). Tissues and cells cultured primarily can be induced, by different ways, to produce proteins, hormones and other metabolites. The cell genome can be altered under experimental conditions in order to achieve research purposes, and different systems of tissue culture and cell manipulation could be assembly to diagnostic techniques and intracellular information flow investigation.

Primary cultures are obtained by disaggregation of a tissue sample, either mechanically or enzymatically. The resulting suspension contains a proportion of cells capable of attaching to a surface, as in tissue culture flasks and dishes, forming a monolayer. Another type of cell culture, the primary explants, consists of cells that migrate out from small fragments of tissue and adhere to the growth surface. Tissue disaggregation is capable of generating larger cultures in short periods of time, but explants culture may still be preferable where only small fragments of tissue are available or the fragility of the cells precludes survival after disaggregation.

Generally, from human beings and large animals, placentae are obtained at term of normal pregnancies and after caesarean section or vaginal delivery. Amnion and chorion membranes are manually separated from each other. The membranes and other placental tissues as well, are washed extensively in PBS or HBSS containing 100 U/ml penicillin and 100 μg/ml streptomycin to avoid bacterial contamination. Blood clots present in the cotyledons may be mechanically removed.

In small animals like rats and mouse (*Rattus norvegicus* and *Mus musculus*), it is a common practice to excise the uteri from pregnant animals, transfer them to Petri dishes with cold PBS, and then, collect and pool the placentae in a new Petri dish. Harvested placentae are washed thoroughly with PBS containing antibiotics. Rat amnion may become distinguished as almost transparent, and may be peeled off from chorion and dissected (14). Otherwise, both membranes are processed together in order to obtain the fetal membrane-derived cells.

Samples are cut into small fragments (3 x 3 or 5 x 5 cm) and may be stored in 50 ml vials filled with serum-free and phenol red-free Dulbecco's Modified Eagle Medium (DMEM) in sterile conditions. The number of cells and their viability, either when stored at room temperature and processed within 24 hours or when stored at 4°C and processed within 28 days, are the same as in fresh membranes, as previously reported (15,16). Those fragments may be either used as primary explants or submitted to enzymatic disaggregation.

Small fragments of amniotic membrane may be directly engrafted to a site of injury, as focal ischemic regions in the heart, lung fibrosis, skin wounds, chronic leg ulcers, and liver fibrosis. The benefits of such therapeutic approaches as the ability to promote re-epithelialization have been shown to reduce inflammation and fibrosis and to modulate angiogenesis (16–19).

Explants culture may also be done using fetal placental villi. Bundles of chorionic anchoring (stem) villi are separated from the chorionic plate and spread out using forceps to separate individual villi from each other. The villi are washed multiple times in PBS, cut into small pieces (1 x 1 mm) and cultured intact. When outgrowing cells reach 80% to 90% confluence, the villi pieces remaining intact are carefully removed, transferred into, and subsequently cultured on new plates to get a sufficient amount of cells for the experimental procedures. Recently it was reported that abundant vasculature present in the human placenta can serve as a source of myogenic cells able to migrate within dystrophic muscle and regenerate myofibers (20).

Instead of using an entire fragment, fetal membranes may be subjected to enzymatic disaggregation, as reported in literature by various different protocols, which differ from each other mainly regarding the enzyme employed and the digestion length. However, all of them include an incubation step at 37°C with some enzyme (dispase, trypsin/EDTA, collagenase) followed by collection of the cells through filtration and centrifugation. Different strategies employing diverse enzymes will be presented. Moreover, it is important to have in mind the kind of cells going to be isolated and the possibility of separating the fetal membranes. Chorion, cotyledons, and especially the amnion are rich sources of mesenchymal stromal/stem cells. The amnion is also comprised of an epithelial layer from which amnion epithelial cells are obtained.

In order to isolate the amniotic mesenchymal stromal cells, first incubate membrane fragments with dispase (2.4 U/ml in PBS) for 9 minutes, followed by collagenase I (0.75 mg/ml) + DNAse (20 μg/ml) solution in PBS for 150 minutes. An alternative option is only one digestion step using 1 mg/ml collagenase I for 120 minutes. Chorionic mesenchymal stromal cells may be isolated by performing two 9-minute dispase digestions, separated by one 9-minute wash step in RPMI-1640 containing 100 U/ml penicillin, 100 μg/ml streptomycin, and 10% heat-inactivated fetal bovine serum (FBS). Another way is to replace the dispase of the first digestion step by collagenase IV. After gentle centrifugation (150 x g for 3 minutes), filtrate each supernatant containing amniotic or chorionic mesenchymal stromal cells in a 100 μm cell strainer. However, mesenchymal stromal cells may be isolated from the placenta without previously separation of its components. In such situation, mince pieces from fetal membranes and incubate for 10 minutes in DMEM with 0.25% trypsin/EDTA, 10 U/ml DNase I, and 0.1% collagenase. Tissues should be pipetted vigorously up and down, avoiding foam, for 5 minutes. Allow large pieces of tissue to settle under gravity for 5 minutes. Transfer the supernatant to a fresh tube, neutralize with FBS, and then spin at 1500 rpm for 10 minutes.

Amniotic undigested pieces (when amnion is incubated either with dispase and collagenase I or collagenase I only, as previously described) may be submitted to another digestion step with trypsin (0.25% in PBS) for 2 minutes, in order to obtain epithelial cells.

Moreover, epithelial cells may be harvested from amniotic membrane fragments after digestion only with trypsin/EDTA, at 37°C. One way is to use the enzyme at 0.125% and incubate three times at 20 minutes each. Alternatively, two digestion steps of 30 minutes each, shaking (0.20% trypsin/EDTA) or two steps of 20 minutes (0.25% trypsin/EDTA) are considered. Trypsin should be inactivated by adding FBS. Collect the cells after centrifugation (1000 rpm for 5 minutes). It is important to highlight that, in all situations, the amniotic epithelial cell layer must be scraped out to remove the underlying tissues, such as the spongy and fibroblast layers, to obtain a pure epithelial layer.

In addition to all disaggregation protocols for amniotic membrane, to achieve different populations of cells (amniotic mesenchymal stromal and amniotic epithelial), the whole membrane may be digested giving a mix of cells, known as amniotic derived cells. Membrane fragments are digested with 0.25% trypsin/EDTA for 30 minutes at 37°C. Cells are collected after filtration in 100 μm cell strainer and three wash steps in cold PBS.

Another protocol requires three digestion steps (1x15 minutes; 2x30 minutes) in 0.05% trypsin at 37°C. Wash cells three times in PBS and collect after centrifugation (1000 rpm for 5 minutes).

Another placental tissue used for cell isolation is cotyledon. Minced cotyledon is digested with 0.25% trypsin for 60 minutes at 37°C. Incubate the undigested fragments with 12.5 U/ml collagenase I for 60 minutes at 37°C. Collect cells by filtration (100 μm cell strainer) and centrifugation (300 x g for 10 minutes).

Finally, when fetal membranes are difficult to distinguish (i.e. small animals as mouse) incubate the placenta with dispase (2.4 U/ml in PBS) for 5 minutes at 37°C, followed by collagenase (0.75 mg/ml) + DNAse (20 μg/ml) in PBS for 90 minutes at 37°C. After filtration (100 μm cell strainer) and centrifugation (300 x g for 5 minutes) a mix of amniotic and chorionic cells are obtained. Alternatively, digest placenta fragments with 300 U/ml collagenase II for 1 hour at 37°C in water shaker. Neutralize enzyme activity with α-minimal essential medium (α-MEM) containing 10% FBS. Collect the cells after filtration and centrifugation.

Cells obtained after any of the procedures mentioned above must be plate onto noncoated tissue culture dishes or flasks. Cells may be seeded at 7x10³/cm² (amniotic mesenchymal stromal cells) and 14-21x10³/cm² (amniotic epithelial cells). The medium should be supplemented with serum, mainly FBS. There is a common practice, in some research centers, to use 20% of serum for the primary culture, and then, replace it for 10% in further passages. However, such cells are perfectly capable to survive and proliferate when incubate with 10% serum from the first passage. The most commonly used media are Roswell Park Memorial Institute (RPMI)-1640, DMEM, α-MEM, and DMEM/F12. It is also important to supplement the medium with antibiotics (100 U/ml penicillin and 100 μg/ml streptomycin or 1% antibiotic-antimycotic solution), L-glutamine (2 mM), nonessential amino acids (1%), β-mercaptoethanol (55 μM), and sodium pyruvate (1 mM). Cells are incubated at 37°C in a humid atmosphere (5% CO_2).

Morphological characterization of the cells, by light microscopy, indicates that amniotic epithelial cells show ground or oval shapes, cluster formation, relatively big nucleus (some can reach half of the cell diameter) and many fat drops in the cytoplasm. Both amniotic and chorionic mesenchymal stromal cells show the fibroblast-like appearance and very high proliferation rate, reaching 70-80% confluence after one to three weeks from harvesting.

Cells may either be used in experimental procedures just right after culturing or be frozen in 10% dimethyl sulfoxide (DMSO) and 90% FBS. An alternative freezing media is 10% DMSO, 40% DMEM, 50% FBS. Prior to using these cells, it is recommended to analyze the viability, which may be determined by use of trypan blue dye, and cell number in a hemocytometer.

In addition, it is important to highlight that all experimental procedures must have ethics committee's approval and, regarding human placental tissues, the previous donor consent.

3.2. Established (transformed) cell lines

There are a lot of cell lines derived from placenta. These cells are cultured with subtle differences, for research purposes and some of them are able to secret hormones and other molecules. Some examples will be present.

BeWo (ATCC® number CCL-98™) is a human epithelial cell line, derived from choriocarcinoma, that produces progesterone, human chorionic gonadotropin (hCG), placental lactogen, estrogen and other reproductive hormones. BeWo is cultured in F-12K medium enriched with 10% FBS. This cell is sensible to human poliovirus 3 and the vesicular stomatitis virus. Other human epithelial cell lines derived from choriocarcinoma are JEG-3 (ATCC® number HTB-36™) and JAR (ATCC® number HTB-144™). JEG-3 produces hCG, progesterone and placental lactogen. The cells are able to transform steroid precursors to estrone and estradiol. The line is cultured with EMEM supplemented with 10% FBS. JAR cells produce estrogen, progesterone, hCG, and placental lactogen and are cultured in RPMI 1640 supplemented with 10% FBS.

The *M. musculus* mast cells 10P2 (ATCC® number CRL-2034™) and 10P12 (ATCC® number CRL-2036™) were established by transformation of placental cells from a 10 day mouse embryo, and are cultured in suspension with RPMI 1640 medium supplemented with 0.05 mM 2-mercaptoethanol and 10% FBS. These cells possess receptors for IgE. Both lines represent a transformed stage of a very early hematopoietic precursor and share surface antigens with multipotent stem cells. The 11P0-1 (ATCC® number CRL-2037™) cell line is similar with the others except for being derived from a 11 day mouse embryo.

The FC-47 (ATCC® number CRL-6094™) is a fibroblast derived from the cat (*Felis catus*) normal placenta, with adherent properties and cultured in DMEM supplemented with 10% FBS.

The 3A-(tPA-30-1) (ATCC® number CRL-1583™) is a human SV40 transformed cell line, derived from placenta, showing an epithelial morphology. When incubate at 40°C, these cells are able to synthesize hCG and alkaline phosphatase. The cells express the transformed phenotype at the permissive temperature (33°C) and the non- transformed phenotype at the non- permissive temperature (40°C). The line has a limited life expectancy of 15 to 18 passages before entering the crisis stage. Medium required for incubation is α-MEM supplemented with 10% FBS.

ChaGo-K-1 (ATCC® number HTB-168™) is a human epithelial cell line, derived from bronchogenic carcinoma, that produces hCG alpha subunit only, estradiol, progesterone, and mucin (apomucin, MUC-1, MUC-2). The cells are cultured in RPMI 1640 supplemented with 10% FBS.

There are several cell lines derived from normal human placenta, which are available for research purposes, such as Hs 726.PI (ATCC® number CRL-7460™), Hs 730.PI (ATCC® number CRL-7464™), Hs 795.PI (ATCC® number CRL-7526™), Hs 798.PI (ATCC® number

CRL-7529™), Hs 799.PI (ATCC® number CRL-7530™), Hs 815.PI (ATCC® number CRL-7548™). All of them are cultured in DMEM supplemented with 10% FBS, at 37°C, in humidity atmosphere (5% CO_2).

3.3. Placenta-derived stem/progenitor cells

Various undifferentiated stem cell sources have been proposed for regenerative medicine, each having their advantages and drawbacks. Embryonic stem cells (ESCs) are characterized by pluripotency and an unlimited self-renewal capacity, but they may present high tumorigenic potential and their use is associated with major ethical concerns. In contrast, mesenchymal stem cells (MSCs) are not ethically restricted, but show limited capacity to proliferate and differentiate into different cell lineages (multipotency). MSCs reside within bone marrow, adipose tissue, dental pulp, and many other tissues. Although bone marrow MSCs are the most studied and the best established, there are some limitations of their use due to invasive nature of bone marrow aspiration, donor site morbidity, inadequate cell numbers, and the limited capacity of proliferation and differentiation. Thus, researchers started looking for alternative sources of MSCs that can be obtained noninvasively and in sufficient quantity. Studies have shown that MSCs may be isolated from placenta (either from maternal or fetal components, especially from the amnion), which are generally discarded as medical waste after delivery, and are therefore without ethical concerns associated with their use. Placenta-derived MSCs have been described to combine characteristics from both embryonic and mesenchymal stem cells: the ability to differentiate into all three germ layers and lack of tumorigenicity. Although specific fetal membrane components have already been considered in the above section "primary culture" of this chapter, here aspects of placenta MSCs will be discussed, without distinction between any fetal membranes as a specific source of cells, even though the amniotic MSCs are the most studied. It should be highlighted that efforts toward standardization of the terminologies used have been made (21) in the First International Workshop on placenta-derived stem cells, who proposed the following nomenclature: aminiotic epithelial cells, aminiotic mesenchymal stromal cells, chorionic mesenchymal stromal cells, and chorionic trophoblastic cells.

One way to identify a population of cells is by performing its immunophenotype characterization, mainly for surface marker expression. Current methodologies employed are immunocytochemistry, immunofluorescence, and flow cytometry. Placenta MSCs reveal profiles in between embryonic and adult stem cells, as summarized in Table 1 where it is shown some of the most used surface markers, as cluster of differentiation (CD), for categorizing stem cell populations. It is important to note that neither embryonic nor mesenchymal cells (and placenta-derived MSC as well) express the hematopoietic stem cell markers CD11b, CD34, and CD45. Placenta MSCs show a positive expression profile for CD29, CD73, CD166, and major histocompatibility complex (MHC) class I molecules (human leucocyte antigen (HLA) -A, -B, -C) and a negative expression profile for CD11b, CD31 (endothelial marker), CD34, CD45, and MHC class II molecules (HLA-DR, -DP, -DQ).

Although there are common properties between placenta and bone marrow derived MSCs, flow cytometric analyses performed by different research groups report that placenta MSCs do not express CD271 (22) and show different patterns for CD90 expression: either low [4.7% (14)], moderate [22.5% (23)], or high expression [> 95% (22)] and even the presence of two distinct subpopulations of placenta MSCs positive for CD90 (24).

Taken together, placenta MSCs satisfy the minimal criteria for identifying multipotent mesenchymal stromal cells (27). Furthermore placenta MSCs, different from other mesenchymal cells, are able to differentiate into ectodermal (neural and retinal cells), endodermal (pancreatic beta cells), and mesodermal (adipocytes, osteocytes, and chondrocytes along with myotubule formation and endothelial cells) lineages *in vitro* (24,25).

Moreover, placenta MSCs are positive for surface markers that are expressed by embryonic stem cells but not by mesenchymal cells. Among them are octamer binding protein 4 (Oct-4), Sox2, Nanog, stage-specific embryonic antigen (SSEA)-1, SSEA-4, GCTM2, Tra-1-60, and Tra-1-80 which are routinely evaluated by immunocytochemistry or flow cytometry (14,23,26). Even when analyzing such expression using more sensitive assays, as in real time polymerase chain reaction (q-PCR), it is observed that placenta MSCs significantly express high levels of Oct-4, Nanog and Sox2 (25). In addition, it was detected by immunofluorescence analysis that human amniotic stem cells, different from other MSCs, are positive for the neural stem cell markers Nestin, Vimentin, Musashi-1, and PSA-NCAM (28). However, all of these data should be analyzed carefully. Some studies indicate variable percentages of positive cells for different pluripotent markers, indicating that more studies are need.

Besides the pluripotency potential of placenta MSCs and the lack of ethical concerns, an important characteristic that lead placenta MSCs to a great importance for clinical application is their low immunogenicity. This property is partially explained by the fact that placenta MSCs do not express MHC class II molecules (HLA-DR, -DP, -DQ) and co-stimulatory molecules *in vitro*, as reported by different research groups and using different methodologies (flow cytometry, qPCR, western blotting), indicating that such cells can be effectively used for both autologous and allogenic transplantations. It was recently demonstrated that fetal membrane MSCs are capable of suppressing proliferation in a mix lymphocyte reaction, decreasing interferon (IFN)-γ and interleukin (IL)-17 production, stimulating IL-10 secretion, and increasing levels of adhesion markers CD54, CD29, and CD49d (29). In addition, placenta MSCs are able to engraft and survive long-term in various organs and tissues without evidence of inflammation or rejection after transplantation into neonatal animals and after in utero transplantation into pregnant rats (30,31). Therefore, placenta MSCs are viable candidates for cell-based therapeutic approaches.

Surface marker	Cell types			References
	ESC	MSC	Placenta – MSC	
CD11b	-	-	-	(24)
CD34	-	-	-	(14,22–27)
CD45	-	-	-	(14,22–24,26,27)
CD29	-	+	+	(14,22,24–27)
CD73	-	+	+	(22–26)
CD90	-	+	ND	(14,22–25,27)
CD105	-	+	-	(23,25,27)
CD166	-	+	+	(23,26)
CD271	-	+	-	(22)
Nanog	+	-	+	(25,26)
Oct-4	+	-	+	(14,23,25,26)
Sox-2	+	-	+	(23,25,26)
SSEA-1	+	-	+	(14)
SSEA-4	+	-	+	(23,26)
GCTM-2	+	-	+	(23,26)
Tra-1-60	+	-	+	(26)
Tra-1-80	+	-	+	(26)

(+): presence; (-): absence; ND: not defined

Table 1. Surface marker expression from embryonic stem cells (ESC), adult mesenchymal stem cells (MSC), and placenta derived mesenchymal stem cells (placenta-MSC).

When considering cellular therapy it is important to keep in mind that a great number of cells will be needed and cell banking may be required. As a result, it is imperative that cells possess some important characteristics: abundant source, high proliferation rate, stability over further passages, efficient freezing, and high viability after thawing. Placenta MSCs have been demonstrated to fulfill such requirements. It was recently reported that placenta MSCs at passage 30 are still able to proliferate in normal rates and keep a stable karyotype (25). Such characteristics are observed in ESCs but not in other types of MSCs, such as those derived from bone marrow or adipose tissue, which don't proliferate well and go to senescence around passages 8-10. Compared to ESCs, MSCs are more resistant to cryopreservation. Whilst ESCs require complex and expansive reagents to be frozen and great number of cells die, placenta MSCs are successfully frozen in 40% DMEM, 50% FBS, 10% DMSO, as well as bone marrow and adipose tissue MSCs, with high recovery rates after thawing. Among many different assays to evaluate cellular viability, the measurement of mitochondrial metabolic rate using MTT (3-[4, 5-dimethyl-2-thiazolyl]-2, 5-diphenyl-2H-tetrazolium bromide) to indirectly reflect viable cell numbers has been widely applied. However, if the purpose is to evaluate the viability and the number of cells immediately before their transplantation to human or animals, trypan blue dye exclusion assay is preferable because only the unavailable cells are dyed, while the desired cells are maintained viable, ready to use.

Before considering using placenta MSCs for clinical purposes, preclinical studies may be performed and the results must be carefully interpreted in order to assure safety for the patient and success in the therapy. Many preclinical studies on placenta-derived cells and amniotic membrane were reviewed (32), considering a wide range of diseases as neurological, pancreatic, muscle, vascular, cardiac, and pulmonary disorders along with liver fibrosis and applications for tissue engineering. Overall the results are promising and, although differentiation of placenta MSCs to specific lineages has been considered the first necessity for therapeutic applications *in vivo*, it seems that the beneficial effects reside on paracrine effects. However, both mechanisms (differentiation or paracrine) are not mutually exclusive and can account for the promising results reported in the literature.

As placenta MSCs share with ESCs some common characteristics such as surface marker expression and plasticity *in vitro*, it was assumed that placenta cells could regenerate damaged tissues due to pluripotency potential. To date, however, preclinical studies have failed to demonstrate the differentiation of engrafted cells. Most of the time the success of the cell transplantation may more likely be due to the secretion of bioactive molecules that could act on other cells and on the microenvironment which they occupy thereby promoting endogenous tissue repair or eliciting other beneficial effects (anti-inflammatory, anti-scarring, angiogenic effects) through paracrine actions.

The abundance of pluripotent cells, high yield, rapid proliferation rates, stable karyotype, plasticity and immunomodulatory properties make placenta MSCs an ideal choice for clinical and tissue engineering applications. Nevertheless, further studies are needed to demonstrate the pluripotency ability of placenta MSCs *in vivo* and to elucidate the mechanisms by which these cells promote physiological and clinical improvements.

4. Conclusion

Eutherian mammalian fetus growth is characterized by the early development of fetal membranes, specially the trophoblast, which produces hormones for the establishment, maintenance and end of pregnancy. The fetomaternal interface varies among species and is a basis for placental structure classification. Primary cultures from placenta are obtained in an easy and not consuming fashion allowing a great number of research applications. Various undifferentiated stem cell sources have been proposed for regenerative medicine. In this setting placenta-derived cells show some advantages as potential to differentiate into the three germ layers, lack of tumorigenicity, low immunogenicity, and more importantly that they are isolated in high yields from sources generally considered medical waste, avoiding ethical concerns.

Author details

João Bosco Barreto Filho*
Federal University of Lavras; Veterinary Medicine Department; Lavras, MG, Brazil

Maira Souza Oliveira
Federal University of Minas Gerais; Veterinary Clinical and Surgery Department; Belo Horizonte, MG, Brazil

5. References

[1] Springer MS, Murphy WJ. Mammalian evolution and biomedicine: new views from phylogeny. Biological reviews of the Cambridge Philosophical Society [Internet]. 2007 Aug [cited 2012 May 1]; 82(3):375–92. Available from: http://www.ncbi.nlm.nih.gov/pubmed/17624960

[2] Carter AM. Evolution of the placenta and fetal membranes seen in the light of molecular phylogenetics. Placenta [Internet]. 2001 Nov [cited 2012 May 10];22(10):800–7. Available from: http://dx.doi.org/10.1053/plac.2001.0739

[3] Liu FG, Miyamoto MM, Freire NP, Ong PQ, Tennant MR, Young TS, et al. Molecular and morphological supertrees for eutherian (placental) mammals. Science [Internet]. 2001 Mar 2 [cited 2012 Mar 22];291(5509):1786–9. Available from: http://www.sciencemag.org/content/291/5509/1786.abstract

[4] Carter AM, Enders AC. Comparative aspects of trophoblast development and placentation. Reproductive biology and endocrinology®: RB&E [Internet]. 2004 Jul 5 [cited 2012 Apr 18];2(1):46. Available from: http://www.rbej.com/content/2/1/46

[5] Mossman HW. Comparative morphogenesis of the fetal membranes and accessory uterine structures. Contrib Embryol. 1937;26:129–246.

[6] Wooding P, Burton G. Comparative placentation: Structures, functions and evolution. XII. Springer; 2008.

[7] Moffett A, Loke C. Immunology of placentation in eutherian mammals. Nature reviews. Immunology [Internet]. 2006 Aug [cited 2012 Mar 19];6(8):584–94. Available from: http://www.ncbi.nlm.nih.gov/pubmed/16868549

[8] Pfeffer PL, Pearton DJ. Trophoblast development. Reproduction [Internet]. 2012 Mar [cited 2012 May 10]; 143(3):231–46. Available from: http://www.ncbi.nlm.nih.gov/pubmed/22223687

[9] Wang Y, Zhao S. Vascular biology of the placenta. San Rafael (CA): Morgan & Claypool Life Sciences; 2010.

[10] Grosser O. Frühentwicklung, Eihautbildung und Placentation des Menschen und der Säugetiere. München: J.F. Bergmann; 1927.

[11] Amoroso EC. Placentation. In: Parkes AS, editor. Marshall's Physiology of reproduction. London: Longmans Green; 1952. p. 127–309.

---

* Corresponding Author

[12] Roberts RM, Chen Y, Ezashi T, Walker AM. Interferons and the maternal–conceptus dialog in mammals. Seminars in Cell & Developmental Biology [Internet]. 2008 Apr [cited 2012 May 3];19(2):170–7. Available from: http://dx.doi.org/10.1016/j.semcdb.2007.10.007

[13] Bazer FW, Burghardt RC, Johnson GA, Spencer TE, Wu G. Interferons and progesterone for establishment and maintenance of pregnancy: interactions among novel cell signaling pathways. Reproductive Biology2. 2008;8(3):179–211.

[14] Fujimoto KL, Miki T, Liu LJ, Hashizume R, Strom SC, Wagner WR, et al. Naive rat amnion-derived cell transplantation improved left ventricular function and reduced myocardial scar of postinfarcted heart. Cell transplantation [Internet]. 2009 Jan;18(4):477–86. Available from: http://www.ncbi.nlm.nih.gov/pubmed/19622235

[15] Hennerbichler S, Reichl B, Pleiner D, Gabriel C, Eibl J, Redl H. The influence of various storage conditions on cell viability in amniotic membrane. Cell and tissue banking [Internet]. 2007 Jan [cited 2012 Mar 10];8(1):1–8. Available from: http://www.springerlink.com/content/326353315g27t273/

[16] Cargnoni A, Di Marcello M, Campagnol M, Nassuato C, Albertini A, Parolini O. Amniotic membrane patching promotes ischemic rat heart repair. Cell transplantation [Internet]. 2009 Jan [cited 2012 Apr 13];18(10):1147–59. Available from: http://www.ncbi.nlm.nih.gov/pubmed/19650976

[17] Cargnoni A, Gibelli L, Tosini A, Signoroni PB, Nassuato C, Arienti D, et al. Transplantation of allogeneic and xenogeneic placenta-derived cells reduces bleomycin-induced lung fibrosis. Cell transplantation [Internet]. 2009 Jan;18(4):405–22. Available from: http://www.ncbi.nlm.nih.gov/pubmed/19622228

[18] Parolini O, Soncini M, Evangelista M, Schmidt D. Amniotic membrane and amniotic fluid-derived cells: potential tools for regenerative medicine? Regenerative medicine [Internet]. 2009 Mar 25 [cited 2012 May 10];4(2):275–91. Available from: http://www.futuremedicine.com/doi/abs/10.2217/17460751.4.2.275?journalCode=rme

[19] Sant'Anna LB, Cargnoni A, Ressel L, Vanosi G, Parolini O. Amniotic membrane application reduces liver fibrosis in a bile duct ligation rat model. Cell transplantation [Internet]. 2011 Jan [cited 2012 Apr 13];20(3):441–53. Available from: http://www.ncbi.nlm.nih.gov/pubmed/20719087

[20] Park TS, Gavina M, Chen C-W, Sun B, Teng P-N, Huard J, et al. Placental perivascular cells for human muscle regeneration. Stem cells and development [Internet]. 2011 Mar;20(3):451–63. Available from: http://www.pubmedcentral.nih.gov/articlerender.fcgi?artid=3120979&tool=pmcentrez& rendertype=abstract

[21] Parolini O, Alviano F, Bagnara GP, Bilic G, Bühring H-J, Evangelista M, et al. Concise review: isolation and characterization of cells from human term placenta: outcome of the first international Workshop on Placenta Derived Stem Cells. Stem cells [Internet]. 2008 Feb [cited 2012 Mar 10];26(2):300–11. Available from: http://www.ncbi.nlm.nih.gov/pubmed/17975221

[22] Kranz A, Wagner D-C, Kamprad M, Scholz M, Schmidt UR, Nitzsche F, et al. Transplantation of placenta-derived mesenchymal stromal cells upon experimental

stroke in rats. Brain research [Internet]. 2010 Feb 22 [cited 2012 Apr 13];1315:128–36. Available from: http://www.ncbi.nlm.nih.gov/pubmed/20004649

[23] Moodley Y, Ilancheran S, Samuel C, Vaghjiani V, Atienza D, Williams ED, et al. Human amnion epithelial cell transplantation abrogates lung fibrosis and augments repair. American journal of respiratory and critical care medicine [Internet]. 2010 Sep 1 [cited 2012 Apr 11];182(5):643–51. Available from: http://www.ncbi.nlm.nih.gov/pubmed/20522792

[24] Ishikane S, Ohnishi S, Yamahara K, Sada M, Harada K, Mishima K, et al. Allogeneic injection of fetal membrane-derived mesenchymal stem cells induces therapeutic angiogenesis in a rat model of hind limb ischemia. Stem cells [Internet]. 2008 Oct [cited 2012 Apr 13];26(10):2625–33. Available from: http://www.ncbi.nlm.nih.gov/pubmed/18669910

[25] Sabapathy V, Ravi S, Srivastava V, Srivastava A, Kumar S. Long-Term Cultured Human Term Placenta-Derived Mesenchymal Stem Cells of Maternal Origin Displays Plasticity. Stem Cells International [Internet]. 2012 [cited 2012 May 8];2012:1–11. Available from: http://www.hindawi.com/journals/sci/2012/174328/

[26] Manuelpillai U, Tchongue J, Lourensz D, Vaghjiani V, Samuel CS, Liu A, et al. Transplantation of human amnion epithelial cells reduces hepatic fibrosis in immunocompetent CCl4-treated mice. Cell transplantation [Internet]. 2010 Jan [cited 2012 Apr 11];19(9):1157–68. Available from: http://www.ncbi.nlm.nih.gov/pubmed/20447339

[27] Dominici M, Le Blanc K, Mueller I, Slaper-Cortenbach I, Marini F, Krause D, et al. Minimal criteria for defining multipotent mesenchymal stromal cells. The International Society for Cellular Therapy position statement. Cytotherapy [Internet]. 2006 Jan [cited 2012 Mar 9];8(4):315–7. Available from: http://www.ncbi.nlm.nih.gov/pubmed/16923606

[28] Kong X-Y, Cai Z, Pan L, Zhang L, Shu J, Dong Y-L, et al. Transplantation of human amniotic cells exerts neuroprotection in MPTP-induced Parkinson disease mice. Brain research [Internet]. 2008 Apr 18 [cited 2012 Apr 13];1205:108–15. Available from: http://www.ncbi.nlm.nih.gov/pubmed/18353283

[29] Karlsson H, Erkers T, Nava S, Ruhm S, Westgren M, Ringdén O. Stromal cells from term fetal membrane are highly suppressive in allogeneic settings in vitro. Clinical and experimental immunology [Internet]. 2012 Mar [cited 2012 May 10];167(3):543–55. Available from: http://www.ncbi.nlm.nih.gov/pubmed/22288598

[30] Bailo M, Soncini M, Vertua E, Signoroni PB, Sanzone S, Lombardi G, et al. Engraftment Potential of Human Amnion and Chorion Cells Derived from Term Placenta. Transplantation [Internet]. 2004 Nov [cited 2012 May 10];78(10):1439–48. Available from: http://content.wkhealth.com/linkback/openurl?sid=WKPTLP:landingpage&an=0000789 0-200411270-00006

[31] Chen C-P, Liu S-H, Huang J-P, Aplin JD, Wu Y-H, Chen P-C, et al. Engraftment potential of human placenta-derived mesenchymal stem cells after in utero transplantation in rats. Human reproduction [Internet]. 2009 Jan 1 [cited 2012 May 10];24(1):154–65. Available from: http://humrep.oxfordjournals.org/cgi/content/abstract/24/1/154

[32] Parolini O, Caruso M. Review: Preclinical studies on placenta-derived cells and amniotic membrane: an update. Placenta [Internet]. 2011 Mar [cited 2012 Apr 12];32 Suppl 2:S186–95. Available from: http://www.ncbi.nlm.nih.gov/pubmed/21251712

Chapter 9

The Art of Animal Cell Culture for Virus Isolation

John A. Lednicky and Diane E. Wyatt

Additional information is available at the end of the chapter

1. Introduction

In virology, cell culture usually refers to the *in vitro* growth and manipulation of cells from a tissue obtained from a multicellular organism. The term "cell culture" is often used interchangeably with "tissue culture". Cell culture remains integral with virology, as viruses are obligate intracellular parasites that require replication within a living cell to produce copies of themselves (i.e., to form progeny virions). Both animal and plant cells are propagated in cell cultures. The only other practical alternatives to cell culture are to propagate the viruses in susceptible animal or plant hosts. This review covers only cell culture for animal viruses. Since the literal meaning of tissue culture is the culturing of tissue pieces, i.e. explant culture, the term "cell culture" is used in this review instead of "tissue culture".

Cell cultures can be prepared from unicellular cells (e.g., white blood cells) or from a piece of "tissue". We define tissue as an aggregate of similar cells forming a definite kind of structural material with a specific function in a multicellular organism. Tissues are first dissociated into smaller pieces by mechanical disruption (such as by cutting into smaller pieces using scissors). Next, the tissue pieces are subjected to treatment with agents that disrupt the extracellular matrix that holds cells together. The cell-dissociating agents usually are proteolytic enzymes such as collagenase and trypsin (to digest proteins) in combination with cation chelators such as ethylenediaminetetraacetic acid (EDTA) that bind, or chelate, the Ca^{2+} and other divalent cations on which cell-cell adhesion depends. The cells are then gently teased apart, suspended in cell growth medium and placed in sterile growth vessels. In the past, glass bottles or petri dishes were used. Indeed, in the context of virology, the term "*in vitro*" originally referred to cells grown "in glass"; this term was used to contrast them experiments carried out on cells grown in glass vessels as opposed to experiments using living organisms. Nowadays, polystyrene (plastic) vessels are most commonly used for cell culture. A majority of cell culture is still performed using techniques wherein the cells are grown in two dimensions. Newer technologies are available that permit cell growth in three dimensions;

using such techniques, it has been able to induce some cells to differentiate into into forms that are not observed during two-dimensional growth. Such technology is still relatively new in the field of virology, and are not discussed in this review.

Since tissues are generally composed of different cell types, a heterogeneous population of cells is usually isolated during the first ("primary") attempt to isolate cells for cell culture. Therefore, subsequent efforts are made to separate the cells into the various types, with the goal of obtaining homogenous populations of cells. The genetic uniformity of a batch of cells can be attained through a process termed "cloning", wherein one cell is isolated and allowed to proliferate to form a "colony" or "clone". All the cells in a colony derive from a common ancestor, and are thus "clones" of each other.

Cells obtained from tissues tend to be "adherent", meaning they attach to the growing surface (of the flask or other vessel) then spread out to form a monolayer. On the other hand, white blood cells settle but do not adhere, and are maintained as "suspension" cultures. As such, they are usually constantly stirred by a spinning magnet in a type of growth vessel termed "spinner flask". Adherent cells growing as flat monolayers are generally of two types: fibroblast and epithelial. Fibroblasts are the most common cells of connective tissue, and synthesize collagen and the extracellular matrix. Epithelial cells line the cavities and surfaces of structures throughout a body, and also form many glands. In culture, they tend to appear rectangular and pack into tight monolayers that look like "brick pavements".

The art of cell culture for virus isolation has entered a renaissance in recent years. Significant improvements have been made in the quality of available reagents, plasticware, and in basic methodologies. For example, disposable platicware has largely supplanted the use of glass vessels. This has reduced the costs of culturing cells in many ways: (a) glass bottles do not have to be washed and sterilized between uses, (b) cell culture plasticware can be purchased ready-made with growth surface coatings or electrostatic "treatments" that promote cell attachment to the growing surface, (c) plasticware is inherently safer in that it is relatively shatterproof. Other sterile disposable plasticware has significantly reduced work burdens. For example, disposable pipettes have made obsolete the task of washing glass pipettes, and the subsequent tasks necessary to prepare them for virology work. In the past, such tasks required the use of multiple detergent and acid washes, which created potential biological and chemical hazards, and required large amounts of distilled water. Apart from improvements in materials used for virology work, many new instruments are available that simplify cell culture technology and at the same time improve precision between experiments. For example, various cell counters are now available that make cell counting easy and reproducible in a intra- and inter-laboratory manner. Moreover, various new or engineered cell lines and primary cells are available for the propagation of viruses once considered very difficult to study *in vitro*. Unfortunately, along with material improvements and technological progress, less emphasis is usually placed on teaching the art of cell culture compared to molecular technology relevant to virus detection. This is a recipe for disaster, as cell culture is neither simple nor a thoughtless process. To paraphrase recent statements by a colleague, "All cells cultured *in vitro* are angry; they are outside of their normal environment and maintained under artificial conditions, surrounded by physiologically incorrect concentrations of all things important to their well-being. No wonder it is so difficult to have well-behaved cell cultures"! Without adequate training and preparation, cell culture as an art and science becomes sloppy, and data generated by such practices are questionable. We have often not been able to repeat the results of others, and they have not been able to repeat ours, due to a difference in the cells used in our experiments. In some cases, the cells are not what they should be, in other cases, the cells are contaminated with adventitious agents that confound the results, and sometimes, the cells have changed, either through differentiation or genetic instability.

Due to the number of different disciplines now engaged in cell culture, the terminology used to describe the work varies substantially. This makes it difficult to communicate effectively using language salient to virology. For this review, eight key terms used for cell culture work and their definitions are presented in Table 1. The definitions given in Table 1 are from the Society for In Vitro Biology (SIVB), as in reference [1]. We strongly recommend adaption of SIVB terminology for inter-laboratory communications of cell culture work. Their definitions are well-thought out, and intuitively understandable.

In this chapter, we provide a review of some contemporary cell culture issues relevant to virus isolation. Some practical guidelines for virus isolation and the maintenance of cell lines are provided. The information we present should provide useful insights for virologists, and may be a useful review of some forgotten principles of virus isolation. This review is not meant to be comprehensive, as each topic would require substantial and detailed treatment, historic and contemporary. It should be noted that the terms "cell strain" and "cell line" (Table 1) are sometimes used interchangeably by other authors. Others also define "cell strain" as a culture of a single type of cell, and "cell line" as an *immortalized* culture of a single type of cell.

2. Basic validation of animal cells used for the isolation of viruses

Cell lines or primary cell cultures derived from vertebrates and invertebrates are used for the isolation of animal viruses. Animal cell lines can be purchased from well-known suppliers, including those listed in Table 2, and primary cells from suppliers such as those listed in Table 3. Alternatively, primary cell cultures can be established *de novo* from live animal cells, tissues or organs. Cells for virology work can also be obtained from university collections or from individual laboratories. In general, the best practice is to obtain cells at the lowest possible population doubling level (Table 1) from reputable suppliers that can provide documentation relevant to traceability. A problem we encounter repeatedly is that a majority of academic research laboratories totally lack or do not have an adequate "Quality System" or "Quality Practices" process, and traceability is problematic. For example, it has been nearly impossible to estimate the population doubling level ("true age of the culture") of cells obtained from most research laboratories due to inconsistencies not only in record keeping but also due to lack of standardization of practices. The major problem encountered with cells that have been highly passaged is that their phenotype can be quite different from their progenitors. This often makes it difficult to reproduce experiments performed in past years using the cells.

Terms	Definitions[a]
Adventitious	Agents which contaminate cell cultures.
Cell line	A cell line arises from a primary culture at the time of the first successful subculture. The term cell line implies that cultures from it consist of lineages of cells originally present in the primary culture. The terms finite or continuous are used as prefixes if the status of the culture is known. The term "continuous line" replaces the term "established line".
Cell strain	A cell strain is derived either from a primary culture or a cell line by the selection or cloning of cells having specific properties or markers. In describing a cell strain, its specific features must be defined. The terms *finite* or *continuous* are to be used as prefixes if the status of the culture is known. If not, the term *strain* will suffice.
Passage	The transfer or transplantation of cells, with or without dilution, from one culture vessel to another. It is understood that any time cells are transferred from one vessel to another, a certain portion of the cells may be lost and, therefore, dilution of cells, whether deliberate or not, may occur. This term is synonymous with the term "subculture".
Passage number	The number of times the cells m the culture have been subcultured or passaged. In descriptions of this process, the ratio or dilution of the cells should be stated so that the relative cultural age can be ascertained.
Population doubling level	The total number of population doublings of a cell line or strain since its initiation *in vitro*. A formula to use for the calculation of "population doublings" in a single passage is: number of population doublings = Log (N/No) X 3.33 where: N=number of cells in the growth vessel at the end of a period of growth. No=number of cells plated in the growth vessel. It is best to use the number of viable cells or number of attached cells for this determination. Population doubling level is synonymous with "cumulative population doublings".
Population doubling time	The interval, calculated during the logarithmic phase of growth in which, for example, 1.0×10^6 cells increase to 2.0×10^6 cells. This term is not synonymous with "cell generation time".
Primary cell culture	A culture started from cells, tissues or organs taken directly from organisms. A primary culture may be regarded as such until it is successfully subcultured for the first time. It then becomes a "cell line". A primary culture may contain multiple types of cells such as fibroblasts, epithelial, and endothelial cells.

[a]Published in reference [1] and also available though: http://www.sivb.org/edu_terminology.asp.

Table 1. Key cell culture terms.

Acronym	Entity
ATCC	American Type Culture Collection, USA
DSMZ	German Collection of Microorganisms and Cell Cultures, Germany
ECACC	European Collection of Cell Cultures, United Kingdom
ICLC	Interlab Cell Line Collection, Italy
LCRB	Japanese Collection of Research Bioresources, Japan

Table 2. Sources of animal cell lines for virology.

	Company
1	Allcells
2	Asterrand
3	ATCC
4	Cell Applications, Inc.
5	Invitrogen
6	Lifeline Cell Technology
7	Lonza
8	QBM Cell Science
9	ReachBio
10	Science Biosystems
11	ScienCell Research Laboratories

Table 3. Partial list of suppliers of primary cell cultures.

Whereas the terminology differs between laboratories and in different countries, many virologists use terminology coined by the ATCC and refer to continuous (previously termed "established") cells as either: CCL (Certified Cell Lines), CRL (Certified Repository Lines), HB (Hybridomas), TIB (Cell Lines in Tumor Immunology Banks), and HTB (Cell Lines in Tumor Cell Banks). In general, cells designated as CCL by the ATCC are the most thoroughly characterized and have been certified by the National Institutes of Health American Cell Culture Collection Committee [2].

Prior to use for virus isolation or detection, cells should be "validated" for their intended purpose. A valid batch of cells implies they have been tested and confirmed suitable for the isolation or detection of the target virus. Three general questions must be asked and answered during the validation process:

a.   Has authenticity been confirmed (are the cells what they should be)?
b.   Do the cells behave as expected before we use them?
c.   When the task is virus isolation (or propagation): Will a wild-type virus similar to the target virus effectively infect the cells, replicate in them, and form progeny virions? If the task is virus detection: Will a wild-type virus similar to the target virus effectively infect the cells and form a virus-encoded product that can be detected by immunochemistry, PCR, or other relevant methods?

Ideally, each laboratory would have the capability of fully "characterizing" each cell line prior to its use for virus isolation. Nominally, this would include some sort of authentication of host cell species, verification that the cells are free of microbial contaminants, confirmation of cell phenotype and growth kinetics, and importantly, the cells should be tested using a contemporary wild-type isolate of the virus that is targeted for detection or isolation. In practice, in-house cell authentication is usually neither cost-effective nor practical for small diagnostic laboratories. For routine/general virology work, we have found that the following practices to be useful for the validation of cells for virus isolation:

a. Obtain cells from a reputable source or laboratory – Resources such as the ATCC provide a certificate that details authenticity, and that the cells are not contaminated with mycoplasma, bacteria, or fungi. Furthermore, cell lines obtained from the ATCC are currently confirmed for species identity using a Cytochrome C subunit I (COI) PCR assay, and Short Tandem Repeat profiling (STR) [3], a PCR-based DNA profiling method for intraspecies identification. Particular attention should be focused on cells obtained through inter-laboratory transfer, as many are cross-contaminated with different cells or have been misidentified altogether [4]. Our experience is they are also often contaminated with adventitious agents.

b. Use a pay-for-service provider to verify cell identity. Whereas we do not necessarily endorse any, examples of commercial service providers that authenticate animal cells are given in Table 4. Many universities also have excellent cell-authentication services.

1	ATCC
2	Bio-Synthesis, Inc.
3	CellBank Australia
4	DNA Diagnostics Center
5	ECACC
6	Genetica DNA Laboratories, Inc.
7	IDEXX RADIL
8	i Life Discoveries
9	Johns Hopkins Genetic Resources Core Facility
10	LGC Standards
11	MicroSynth
12	Orchid Cellmark
13	SeqWright

Table 4. Partial list of cell-identification service providers.

c. Observe the cell morphology at low and high cell densities, and verify they conform to expectations. Aberrant cell shapes may be an indication that the wrong cells were obtained, or that the cells are contaminated with mycoplasma or other infectious agents.

d. Verify the cells are mycoplasma-free. We have obtained mycoplasma contaminated cells from even the most prestigious laboratories; in one instance, 15/20 cancer cell lines were contaminated, and upon genetic analysis, five different mycoplasma species were uncovered in the cells. Each laboratory should have a mycoplasma test, suitable for the detection of a wide variety of mycoplasma, of which there are over 100 species. Many PCR tests for mycoplasma have been described, and various kits are available commercially. We have found through experience that PCR tests should be carefully evaluated, as many of the older tests do not detect some of the more recently discovered mycoplasma species. In concert with PCR, mycoplasma isolation is recommended as a second test, especially if the PCR tests have not been updated. We specifically recommend culturing cells for a minimum of two weeks in the absence of antibiotics prior to performing mycoplasma tests, as some mycoplasma species are inhibited but not killed by antibiotics added to cell growth media. It is a good practice, however, to isolate any incoming cell lines and automatically treat as if they contain mycoplasma through the incorporation of plasmocin in the growth medium. Some of the other non-PCR based tests for mycoplasma include detection of live organisms using the Barile culture method [5], staining of cell cultures with fluorescent dye Hoeschst 33258 [6], and use of PlasmoTest (InvivoGen), which is a cell-based assay. PlasmoTest is performed using HEK-Blue™ -2 cells, which are HEK293 cells that are engineered with multiple genes from the toll-like receptor 2 (TLR2) pathway. In the presence of mycoplasma, HEK-Blue-2 cells secrete embryonic alkaline phosphatase, which then reacts with a specific substrate in the cell media to produce a blue color.

Apart from university testing facilities, service providers can also perform mycoplasma tests. Whereas we do not necessarily endorse any, examples of service providers that perform mycoplasma tests are listed in Table 5:

	ATCC
1	Bionique Testing Laboratories, Inc.
2	BioOutsource
3	Bioreliance
4	Charles River Laboratories
5	Clongen Laboratories, LLC
6	Invivogen
7	Lonza
8	Microtest Laboratories, Inc.
9	Minerva Biolabs
10	Mycosafe
11	Nucro Technics
12	Paragon Bioservices
13	Q Laboratories, Inc.
14	Quadscience
15	Wuxiapptech

Table 5. Partial list of mycoplasma testing services

e.   Evaluate the cells for contamination with agents other than mycoplasma [7]. Often times, laboratories assume that bacterial contaminants are easy to detect through the color of the pH indicator dye in the growth medium. In particular, many growth media formulations incorporate phenol red, and a rapid conversion from red to yellow is used as evidence for the presence of bacterial contamination. Whereas the former is often true for bacteria that are fermenters or facultative anaerobes, it should be noted that obligate aerobes tend to make the media more basic. Moreover, some bacterial contaminants are missed during cursory inspection by microscopy because they are non-motile, replicate slowly, and are often present attached to the surfaces of the cultured cells. We have found *Propionibacterium acnes* (which is an aerotolerant anaerobe), various actinomycete species, and both gram positive and negative anaerobes in contaminated cell cultures. Often, the bacteria are resistant to antibiotics commonly used for cell culture, and worse, some are multiple-drug resistant and nearly impossible to eradicate.

f.   As a final validation step, verify the cells are suitable for the target virus. This is especially true for cancer cell lines, which are karyotypically abnormal or genetically unstable. Where possible, we examine the susceptibility of the cells to a contemporary, wild-type version of the target virus, as well as to a known laboratory strain of the same virus that serves as a reference standard. This comparison is made because viruses tend to mutate, and in the process, the affinity for the particular cellular receptor may change. Moreover, the quantity of cellular receptors on the cell may change, and this can affect the avidity of the virus for its receptor.

3. Complications arising from the use of primary cells for virus isolation

Primary cells (Table 1), which are non-immortalized cells taken directly from a living organism, are often used in clinical laboratories for the isolation of various viruses. For example, primary monkey kidney cells, which in the USA are obtained from rhesus or other macaques or from various African "green" monkey species, are used for the isolation of echo and other picorna viruses, and human parainfluenza and other paramyxoviruses. Primary cells are especially useful for diagnostic virology because some viruses are easier to isolate (or can only be isolated) in them. However, primary cells often harbor latent viruses that become reactivated once the cells are separated from kidneys and propagated *in vitro*, or, contain viruses that produce a persistent but subclinical infection of the host. The latter viruses may not cause significant (if any) pathology *in vivo*, where the cells exist in an environment with a functional immune system. But outside of the host and away from the immune system, the cells may be fully permissive and the same virus cause highly cytopathic effects (CPE). Unfortunately, some primary cells may also harbor viruses that can replicate in the host cells without causing easily recognized CPE, and also in the indicator cells used for their isolation (or detection) *in vitro*. Unwanted viruses in primary cells cause various complications relevant to the isolation of a target virus, including:

a.   They might quickly overtake a cell culture, reducing the chances of isolating the target virus.

b.   They may cause CPE identical to those of the target virus, thus causing a false positive preliminary assessment.

c.   They are obvious sources of contamination that complicate the isolation of the target virus in "pure" form.

d.   They may pose a biosafety risk to laboratory workers.

Noteworthy, primary cells can harbor contaminating agents other than viruses. For example, mycoplasma species are present in most animals, and are prevalent on the surfaces of the respiratory tract. Moreover, mycoplasma species exist as intracellular and extracellular varieties. For reasons not yet entirely clear, kidneys are "sanctuaries" for viruses. For this reason, we often hunt for new viruses in kidney cells sourced from exotic species (J. Lednicky, unpublished).

Some of the adventitious agents we have encountered in primary cells include:

- *Human cytomegalovirus* in human kidney cells
- *Lymphocytic choriomeningitis virus* in mouse kidney cells
- *Murine polyomavirus* in the kidneys of mice from a university rodent colony
- *Parainfluenza virus* 5 (formerly *Simian virus* 5) in rhesus monkey kidney cells
- *Simian cytomegalovirus* in simian (various species) kidney cells
- Simian foamy retrovirus in rhesus and green monkey kidney cells

Primary cells also have a finite lifespan, and should be used with minimal passages *in vitro*. Otherwise, senescence of the cells can be mistaken for CPE caused by viruses. Various commercial suppliers currently provide primary cells from various tissues and species. These cells should be used judiciously for virus propagation or isolation. A common mistake is to assume that primary cells obtained from the suppliers are certified to be virus free. In reality, this is not the case. For example, the *donors* of primary human cells sold in the USA are examined (by serology) for antibodies to Hepatitis B and C viruses, and to HIV, following United States Food and Drug Administration (USFDA) guidelines, and if that information is not available, the *cells* are checked by PCR or other methods for the same viruses. [The USFDA is an agency of the United States Department of Health and Human Services responsible for protecting and promoting public health through the regulation and supervision of biopharmaceuticals, blood transfusions, cosmetics, dietary supplements, electromagnetic radiation emitting devices (ERED), food safety, medical devices, over-the-counter pharmaceutical drugs (medications), tobacco products, prescription, vaccines and veterinary products]. However, additional tests for other adventitious agents have not been mandated by the USFDA, and it may be impractical to check for the presence of many other agents with regard to cost and representative sampling reasons. Thus, commercially supplied human primary cells are sold with an advisory statement indicating the cells should be considered as potentially infected, and that biosafety practices be used when working with the cells.

Cell deterioration in primary cells due to improper cell growth media formulation can also be confused for CPE caused by viruses. It is important to maintain non-infected controls along with cells used for virus isolation for comparison. We have noted cell deterioration due to L-glutamine deficiency, and to improper dosage of antifungal agents in the growth media, among a few batches of commercially bought primary cells. Similarly, commercial media formulations for primary human cells often include additives such as epinephrine, human recombinant epidermal growth factor, hydrocortisone, insulin, transferrin, and others; a mistake in the amount of some of these biomolecules added to the cell growth media can adversely affect cell viability.

Thus, primary cells are useful for the isolation of some viruses, but should be used with caution because: (a) they can contain adventitious agents, and (b) cell deterioration due to one of many different reasons can be mistaken for virus-induced CPE.

4. Serum vs. serum-free cell culture media

Serum has been a mandatory additive in cell growth media for much of the history of tissue culture, and is essential for cell growth, metabolism, and to stimulate proliferation ("mitogenic effect"). This is because serum in a complex mixture that provides (a) hormonal factors for stimulating cell growth and proliferation and promoting differentiated functions, (b) transport proteins carrying hormones (e.g. transcortin), minerals and trace elements and lipids, (c) attachment and spreading factors, and (d) stabilizing and detoxifying factors needed to maintain pH or to inhibit proteases either directly (e.g., α-antitrypsin inhibitor in serum is an important inhibitor of the protease trypsin), or indirectly, by acting as an unspecific sink for proteases and other (toxic) molecules [8].

Previously, human and horse serum, collected aseptically through venipuncture, was the source of serum for tissue culture. This was later supplanted by less expensive bovine serum sourced from blood taken from slaughterhouse bovines. The bovine blood is collected using somewhat crude methodology, the blood clotted, serum separated, then usually filtered using 0.1 μm filters. Both calf and fetal bovine serum (FBS) are used for cell culture media. However, primarily because of its rich content of growth factors and its low γ-globulin (antibody) content, FBS has been adopted as the standard supplement of cell culture media [8]. Unfortunately, filtration using smaller pore filters is technically difficult due to the complex composition of serum. Moreover, mycoplasma and viruses are not always retained by the filters; mycoplasma presumably due to small size and their inherent flexibility, and viruses due to their small dimensions and often pleomorphic nature. Thus, serum itself has often been the source of mycoplasma and virus contamination [9-14]. Viruses that may be common contaminants of bovine calf or FBS include: bovine viral diarrhea virus (BVDV) [15-19], bovine polyomavirus [20, 21], bovine parvovirus [22-24] (J. Lednicky, unpublished), and bovine herpes viruses [25-28]. Inadvertent contamination of cultured cells by these serum-derived viruses has obvious repercussions not only with regard to data generation, but also because it exerts a toll on time wasted in the performance of laboratory work, and the costs thereof. And it is usually the case that problems are noted many months afterwards (in common language, the problems occur "downstream").

Whenever economically feasible, we suggest using gamma-irradiated low antibody FBS or calf serum for tissue culture media. Low antibody sera are desirable to reduce the chances of antibody neutralization of the target virus, thus improving chances of virus isolation.

Gamma-irradiation is generally performed after filtration, and acts as a secondary safeguard [29-31]. The irradiation process purportedly inactivates viruses and live microorganisms with minimal damage to product integrity [32, 33]. Gamma-irradiation should be performed using a validated process; we have occasionally encountered batches of gamma-irradiated sera replete with filamentous carbonaceous material that to the untrained eye may be mistaken for fungal mycelia. The presence of filamentous material is due to protein degradation resulting from improper handling during the gamma-irradiation process. A word of caution: We have noted that small DNA viruses such as parvo- and polyomaviruses in sera are not effectively inactivated by gamma-irradiation (Lednicky and Wyatt, unpublished observations). Similar observations were recently published by others [34, 35]. Knowledge over the susceptibility of cell lines to bovine parvoviruses and polyomaviruses is relatively scant.

Another important consideration when purchasing calf serum or FBS is the level of endotoxin [36]. Failure of some cell culture attempts can sometimes be traced to the level of endotoxin in the sera. When presented with the choice, it is always best to purchase sera with the lowest possible endotoxin levels. Industry standards for serum sold at present specify < 10.0 EU/ml, where 1 EU/ml ~ 0.1 ng/ml (EU = endotoxin unit). Endotoxin is detected by the limulus amebocyte lysate assay, which can detect down to 0.01 EU/ml.

Risks associated with the use of animal-sourced components in the culture milieu have led to the quest for protein-free, animal-free cell growth media. To-date, various cell lines have been adapted to grow in serum-free media. And numerous defined serum-free media formulations are available for some of the cell lines commonly used for virus isolation, such as for the African green monkey kidney cell line termed Vero, and for Madin Darby Canine Kidney (MDCK) cells. The use of serum-free cell growth media has been validated for the isolation or propagation of many viruses [37], including those used for vaccines. As an example, a chimeric parainfluenza virus type 3 - respiratory syncytial virus was propagated to 100-fold higher titers in Vero cells grown in serum free cell growth media than could be attained with a serum-containing media formulation [38]. Vero cells grown in serum-free media has been used for the production of reovirus [39]. MDCK cells grown in serum-free media was used for the production of Influenza H5N1 virus used as a vaccine [40]. Rabies virus used for vaccine production was grown to higher titers in Vero cells in serum free than in serum-containing cell growth media [41].

Thus, serum free media should be considered for virology applications that entail routine virus propagation or vaccine virus production. Less explored is the use of serum free media for the isolation of viruses from clinical specimens. To-date, many laboratories have not experimented with this option, probably because many serum-free media formulations are used without the addition of antibiotics and antifungals, or with their use at 1/5 to 1/10 the concentrations that would normally be used in serum-containing media. At lower concentrations, the antibiotics and antifungals might not effectively suppress contaminating microorganisms that are often present in clinical specimens. However, we have found most cells that have been adapted to growth in serum-free media do tolerate antibiotics. For example, we have used MDCK cells in serum-free media with antibiotics for the isolation of influenza viruses. As there are now many different serum-free cell media formulations, it is likely that successful methodologies using these for diagnostic virology will be developed. A particularly important aspect of such work would be the reduction of costs for diagnostic virology laboratories in less privileged nations, since serum is expensive.

As pointed out above, serum is a complex mixture and lot to lot variation and inconsistency is common. This is because the source animals themselves differ, and their nutrition status, hydration, over-all wellbeing, etc. have direct effects on the quality of the serum. Hormone and vitamins levels in the sera can vary, and all things taken together, can have a significant impact on cell culture. Thus, another argument for using serum-free cell growth media, where possible, is there is potentially better process control. Prior to the development of serum-free cell growth media, many researchers would test multiple serum lots for a particular application, then purchase a large lot of the best performing batch of serum. This is still an advisable practice when serum must be used for long-term projects, and for high-throughput work, but imposes large cost and storage burdens.

5. Remediation of mycoplasma contamination

Mycoplasma are bacteria (class Mollicutes) that are among the most common and serious contaminants of cell cultures. There are two genera of mycoplasma that are relevant to cell culture, *Acholeplasma* and *Mycoplasma,* and they have several unique properties that distinguish them from other prokaryotes. In particular, they lack a cell wall, instead using sterols to maintain their plasma membrane. *Mycoplasma* require cholesterol or similar sterols derived from vertebrate hosts, which they incorporate into their cell membranes, whereas *Acholeplasma* grows in the absence of sterols (but incorporates them if present). Since they lack cell walls, mycoplasma are unaffected by antibiotics that interfere with murein (peptidoglycan) formation of cell walls, such as penicillin and other beta-lactam antibiotics. They are also resistant to streptomycin. In the early 1990's, it was estimated that about 15% of cell cultures in the USA were contaminated with mycoplasma [42]. The most likely sources of mycoplasma for laboratories engaged in cell culture are: (a) previously contaminated cell cultures (which can include new cultures from unknown sources or some obtained from cell banks), (b) laboratory equipment, media, reagents that came into contact with contaminated cultures, and sera used for cell cultivation [43], and (c) personnel involved in cell maintenance [44].

Because they are relatively small (0.15–0.3 μm), it is difficult to filter them out of suspension. Both intracellular and extracellular types of mycoplasma exist. Mycoplasma replicate relatively slowly as they spread through a cell culture. A few mycoplasma species comprise 95% of cell culture isolates: *Acholeplasma laidlawii, Mycoplasma arginini, M. fermentans, M. hominis, M. hyorhinis, M. orale, M. pirum,* and *M. salivarium* [42, 44-46]. It should be noted that by definition, members of the genus *Mycoplasma* are restricted to vertebrate hosts. For this reason, many researchers assume plant-based materials are free of mycoplasma that infect cells derived from vertebrates. However, we suggest caution against such notions; as contamination of plant-based media components by *A.laidlawii* has recently been reported [47]. The consequences of mycoplasma contamination of cultured cells may vary from subtle to severe. The overall effects of mycoplasma on a cell culture vary according to the mycoplasma species infecting the cells, the mycoplasma burden (titer), the type of cells infected, and the duration of the infection [48]. Mycoplasma attach to cell membranes in order to obtain nutrients, and in the process can damage the host cell's membranes, DNA and RNA, and intracellular organelles. Their presence can exert profound effects ranging from unexpected alterations of growth patterns and host gene expression to modulation of host metabolism (e.g., produce pH-altering metabolites), induction of chromosomal aberrations, depletion of media, alteration of product yields (such as virus titer), and alteration of cytokine production and other functions of cells of the immune system [44, 48-50]. It goes without saying that efforts must be exerted at *preventing* mycoplasma contamination of cell cultures, following good cell culture practices [44].

If a cell culture is contaminated with mycoplasma, there are two remediation solutions: (a) Destroy (autoclave) the culture and dispose of it, and start with a new culture, or (b) Treat the culture with specific antibiotics or other biochemicals that are toxic to mycoplasma but safe for cells. Most commonly used cell culture antibiotics are not effective against mycoplasma contamination but other antibiotics have shown success in eliminating mycoplasma from cell cultures. In the near past, the following treatments were used to clear mycoplasma-contaminated cell cultures: (a) one to two week treatment with the fluoroquinolone Mycoplasma Removal Agent (MRA, from ICN Biochemicals [now MP Biochemicals]), (b) two week treatment with the fluoroquinolone ciprofloxacin (which is better known by many as Ciprobay), and (c) three rounds of sequential one-week treatment with BMCyclin (Roche), which contains a pleuromutilin and a tetracycline derivative [51, 52]. Prior to the use of the aforementioned mycoplasma eradication products, long-term treatments with tetracyclines were common. We and others have found such mycoplasma eradication treatments to be of limited efficacy. Instead, when necessary, we favor the use of plasmocin (InvivoGen) treatment (for a minimum of two weeks). Plasmocin is an antbiotic mixture that consists of a combination of a macrolide and a quinolone, and it is active against both intra- and extracellular mycoplasma. No permanent alterations were detected in eukaryotic cells treated with plasmocin [53], suggesting the product may be generally safe for most cell lines. This has been our experience so far with the many cell lines we have treated upon receipt as a prophylactic measure (J. Lednicky and D. Wyatt, unpublished). Plasmocure, a newer antibiotic mixture for the eradication of mycoplasma from cell cultures, is now also available from Invivogen; however we have not tested this product, and cannot comment on its efficacy.

One last important consideration is some cultures may be infected by more than one mycoplasma species [54; J. Lednicky and D. Wyatt, unpublished data]. Failure to rid a culture of mycoplasma may arise if one (or more) of the species in a contaminated culture is resistant to the antibiotics being used to eradicate the mycoplasma. Because mycoplasma biology varies among the species, it is not correct to generalize the effect of a certain antibiotic or mixture of antibiotics for all (i.e., an antibiotic or antibiotic mix that works for the eradication of one may not work for a different mycoplasma species) [55].

In the last few years, we have not detected mycoplasma in cell lines that we have validated in our respective laboratories. And we have noted a decrease in the percentage of mycoplasma-contaminated cells obtained from reputable laboratories. However, we still frequently detect mycoplasma in virus preparations obtained from commercial suppliers, clinical laboratories, and other sources. Thus, it is advisable to assess virus stocks for mycoplasma contamination. This is particularly important for cell culture-produced virus stocks that are intended for animal studies. Especially so if the virus must be injected into the animals through intracerebral, subcutaneous, or intraperitoneal routes. In animal studies, it is rarely the case that researchers examine whether inter-laboratory discrepancies might be due to the presence or absence of mycoplasma in the challenge virus. Related to this, it is also worthwhile to verify that hybridoma antibodies used for animal studies are mycoplasma free.

6. Microscopy for cell culture

We have noted that many laboratories engaged in cell culture, including virology laboratories, are up to now only equipped with microscopes that use transmitted brightfield illumination. And many of these laboratories use microscope objectives that in combination with the eyepiece lenses produce a magnified image of only 200X or so. A big problem is that cells in culture appear virtually transparent when observed with an optical microscope under brightfield illumination. To improve visibility and contrast, analysts must then reduce the opening size of the substage condenser iris diaphragm, but this usually results in a serious loss of resolution and the introduction of diffraction artifacts. Moreover, subtle morphological changes are impossible to observe when cells are viewed using low-magnification using transmitted brightfield illumination. Thus, it is easy to miss telltale signs of cell deterioration or "stress", and also of CPE, when cultured cells are viewed using transmitted brightfield illumination at relatively low magnification. The images of cells photographed using transmitted brightfield illumination under low magnification tend to lack adequate resolution for teaching purposes and are less desirable for data capture. Indeed, laboratories that utilize transmitted brightfield illumination often resort to the use of various cell-staining procedures so that features of virus-infected cells can be more readily visualized. This adds costs and often presents biosafety hazards, and also kills the cells being studied. When cell staining must be used as an adjunct for microscopy, the following are useful:

1. May-Grüenwald-Giemsa stain, used to visualize cytopathology. Stained infected cell monolayers are compared to those of non-infected controls. The May-Grüenwald-Giemsa stain will differentially stain DNA nucleoproteins (red/purple) and RNA nucleoproteins (blue). The stained cells are also examined for the presence of syncytia or giant cells, inclusion bodies, and other viral CPE.

2. Modified Wright-Giemsa stain for white blood cells, suspension cells, or cells grown as monolayers on coverslips. Available commercially as Diff-Quick test, we have found this staining process useful for the detection of coxiella- or chlamydia-infected cells, viral inclusion bodies, and similar applications. Chromatin margination, cytoplasmic stranding, vacuoles, and the shape of viral inclusion bodies are easily visualized, giving clues on which virus may be in a cell.

3. Gimenez stain. This stain is useful for intracellular bacteria that stain poorly using a Gram stain. We have used this stain to detect bacteria growing attached to the outside of cells grown as monolayers in cell culture.

A better way to visualize cultured cells is to use an inverted microscope that is set up for phase contrast microscopy. With phase contrast, relatively thin objects such as flat cells become visible under the light microscope. Differences in cell structure are amplified during phase contrast, resulting in an image that can be regarded as an optical density map. Fine structures in the cells not easily detected by brightfield microscopy are clearly visualized, facilitating assessment of cell "condition", and CPE are easier to detect. Phase contrast microscopy is especially advantageous because living cells can be observed without being killed, "fixed" (preserved), and stained. We equip our phase contrast microscopes with objectives that provide a wide, flat view, and recommend magnification of up to at least 400X. Economical, high-quality digital cameras are now widely available and highly recommended for image capture. There are various microscopes on the open market that are affordable, rugged, and suitable for most virology applications using cells grown as monolayers. The authors of this manuscript use microscopes made by Leitz, Nikon, and Olympus in their laboratories. For high containment laboratories wherein face shields must be used [e.g., face shield that is a component of a powered air purifying respirator (PAPR) assembly], a focusing screen in LCD format is convenient. Otherwise focusing is difficult, especially when the face-shield is scratched from everyday "wear and tear", or has turned somewhat opaque from repeated decontamination with sodium hypochlorite or similar harsh decontaminating solutions).

Within the last few years, the concept of using digital inverted microscopes has generated much interest among virologists. We have tested the EVOS-x1 brightfield and phase contrast microscope (AMG, Germany) and find it easy to use. These microscopes have a high resolution LCD viewing screen that is useful for instruction purposes, as it can be simultaneously be viewed by many people. These microscopes are portable and have a relatively compact overall dimension that makes them facile to install within biosafety cabinets or class III gloveboxes. Focusing on the image by viewing the LCD screen is convenient in high-containment biosafety laboratories wherein work is performed using protective face masks such as work with PAPR.

7. Maintenance of virus-infected cell cultures over a few months

The isolation of some viruses may take weeks to two months or more. There are many reasons for this. In many cases, the indicator cells are sub-optimal for high-titer replication of the virus. Some viruses associated with lesions may be defective; for example, *Measles virus* (MeV) associated with subacute schlerosing panencephalitis may cause cytopathic effects without liberating virions [56 and references therein]. Other viruses cause persistent, sub-clinical infections in their natural hosts and may have evolved to replicate slowly. Thus, prolonged incubation, often up to two months, is required for the isolation of some polyomavirus JC virus variants that have an archetypal regulatory region [57; J. Lednicky, unpublished] . We have been able to isolate viruses considered "nearly impossible" to isolate *in vitro* by maintaining cell cultures for prolonged periods post-inoculation with clinical (or environmental) samples. A companion process to the long-term maintenance of cultures is the periodic performance of "blind passages" of the infected cell (discussed in more detail below).

The key to maintaining cells over many months is to slow down their metabolism and/or their population doubling time. This is performed on a case-by-case basis, since not all cells will respond the same way to any given protocol, and there may be variability within a given cell line, depending on age, passage history, etc. An important goal is to maintain the cells in a state relevant for the target virus. For example, would the target virus normally infect (and replicate in) cells that are mitotically active/dividing, as for parvoviruses? If so, the cells should not be allowed to become confluent. On the other hand, does the virus normally infect contact-inhibited "confluent" cells? Should the cells be terminally differentiated? Must the cells be rotated to create air and liquid interfaces as performed for the isolation of rhinoviruses using cells in roller bottles?

The three usual ways of maintaining cells for long-term observation are:

a.   Use growth medium with a reduced glucose concentration. For example, commercial Dulbecco's Modified Eagle Medium (DMEM) is sold as "high" or "low" glucose formulations. Cells can be propagated in high glucose DMEM, then transitioned to low glucose DMEM after they are inoculated with virus.

b.   Switch from nutrient rich to traditional growth media formulations after the cells are inoculated with virus. For example, cells might be grown in DMEM, but once infected, maintained in Eagle's Minimum Essential Medium (EMEM). As a rough measure, DMEM has about 4x the concentration of amino acids found in EMEM.

c.   Reduce the amount of serum in the cell growth medium after the cells have been inoculated with virus. For example, instead of 10% FBS, in many cases cell growth media with 1 – 3% serum can be used to re-feed cells. Furthermore, instead of FBS, calf serum can sometimes be used for the maintenance/re-feed media. This helps to slow down cell replication and metabolism because overall, calf serum has a lower concentration of growth promoting factors than FBS.

We have used the procedures above, solely or combinations thereof, for the isolation of various viruses. It is a good idea to validate growth media for each application; we have found that cell growth media are not necessarily equitable between manufacturers (i.e., all things kept equal, DMEM from supplier A may not work as well as DMEM from supplier B).

One technique that that has worked well for the isolation of viruses that infect and replicate slowly in cells derived from African green monkeys is this: substitute Vero E6 for other cell lines derived from African green monkey cells. Vero E6 cells are more contact-inhibited than similar closely related cell lines, and can be maintained for long periods of times with minimal to no media changes. For example, we have been able to isolate human metapneumoviruses (CPE detected in 10 – 14 days) and other paramyxoviruses such as parainfluenza 4A and 4B viruses in Vero E6 cells maintained in serum-free media, with changes of the growth media every two weeks. This concept was popularized by Akibo *et al* [58] for the isolation of metapneumoviruses, whereas Vero E6 cells were previously thought less useful than tertiary monkey kidney (tMK) and the cell line LLC-MK2 for the isolation of those viruses.

8. Adventitious viruses in cell-lines

It is not uncommon to receive virus-contaminated cell lines from suppliers, and this is especially true for cells obtained through inter-laboratory transfer. One problem is that the cells may have become infected with bovine viruses (from serum) that replicate relatively slowly (i.e., the time it takes for them to complete a replication cycle and form progeny virions is higher than that of the cell population doubling time). These contaminating viruses are referred to as "adventitious" viruses (i.e., they are viruses that should not be present).

Many times, the adventitious viruses go unnoticed, and the deterioration of the cells is attributed to some type of "folklore" prevalent among cell culture practitioners. For example, we have encountered batches of cells that deteriorated when seeded at low densities, but not at high densities. Researchers who had been working with those particular cells did not question the cell propagation instructions provided with the cells. We discovered the reason the cells survived when seeded at high densities was because they were infected with parvovirus, and many parvoviruses require actively replicating cells to form progeny virions *in vitro*. Upon further investigation, we found that some cells lines that are available from commercial sources are packaged along with instructions that suggest that for propagation, no more than five cell passages (with seedings at high densities) should be attempted for "optimal" results. We surmise that a similar issue exists for those cells (that they are infected with parvo- or other viruses that require actively replicating cells)!

Apart from sera, contaminating viruses can also be traced to laboratory workers, to animal-sourced enzymes used for cell culture (such as porcine trypsin), and to other biological used for cell culture. A recent compilation of bovine and porcine viruses that may contaminate bovine serum and porcine trypsin is available in ref. 59. As new viruses are discovered, awareness of their possible presence in biologicals like sera and trypsin draws more interest and attention. For example, porcine trypsin has been traced as the source of *Torque teno sus virus* (TTSuV), a member of the family *Anelloviridae* that is a contaminant of many cell lines. Indeed, TTSuV was found in fifteen cell lineages, originating from thirteen different species, and its presence in the cell lines probably traced to the use of porcine trypsin during the propagation of those cells [60]. Anelloviruses are small DNA viruses that replicate within the nuclei of infected cells, and CPE due to their presence have not been well described at present. Porcine circoviruses 1 and 2 (PCV1 and PCV2) have also been detected in cell lines including those used for vaccines, and have been traced to the use of porcine trypsin [61-63].

We recently traced a filtered amino acid supplement as the source of a contaminating reovirus, and learned from some industry colleagues they had made the same finding. However, as typical of these cases, the findings are not published and thus the information not widely disseminated. Reoviruses however can have wide host range and infect many different cell lines [60, 64].

In some cases, unusual bacteria, and even some single-celled eukaryotic microorganisms cause cell contamination problems that are attributed to viruses. This is because many people engaged in cell culture have little experience with the detection and identification of these types of organisms. We were once tasked with identifying a "virus" affecting some important in-house developed cancer cell lines, which turned out to be infected with chlamydia.

Adventitious viruses can confound research results in many unexpected ways. In one memorable event, cells were thought to have been transformed through a "hit and run" mechanism, as the transformed cells did not retain an oncogene that had been transfected into the precursor (non-transformed) cells. However, it was shown that the cells were infected with bovine polyomavirus, and were expressing its tumor protein genes (Lednicky, unpublished observations). This dashed the investigators' hopes for a patent application. It is also distressing when one performs electron microscopy and discovers that more than one virus is present in the specimen being viewed (or worse, only the wrong virus is visualized). Contaminated cell lines are a main reason gene expression studies can vary significantly between laboratories. Biopharma and the vaccine industry are by now very cognizant of the dangers posed by using contaminated cell lines, due to historic and recent problems caused by adventitious viruses.

In recent years, we have helped various researchers as well as biotechnology companies identify adventitious viruses affecting their work. Among the adventitious viruses we have recently found are:

- *Bovine herpesvirus* 4 (BoHV-4) in human HeLa and Hek293, and in canine MDCK cells. BoHV-4 has been reported by others to be capable of infecting cells of various different species, including human cells [65]. In MDCK cells, cells infected with some strains of BoHV-4 produce vacuoles at the cell to cell junctions, and this is often mistaken for "cell stress" due to nutrient depletion.
- BVDV in MDBK (Madin Darby Bovine Kidney) cells. This finding is consistent with the known biology of BVDV, and MDBK cells are often used for *in vitro* studies of BVDV. For example, see [66].
- *Mouse minute virus* (MMV) in Chinese hamster ovary (CHO) cells. MMV is a notorious contaminant of rodent cell lines important for the biopharmaceutical industry [67].

- Reovirus in various mammalian and insect cell lines. Electron microscopy was used to identify these viruses; follow-up tests were not performed to determine the viral species. Nevertheless, the presence of reoviruses in insect cell lines was a surprise to our industry clients, who experienced catastrophic losses of insect cell lines used for baculovirus-based technologies.
- Vesivirus in Crandell–Rees Feline Kidney (CRFK) cells. We have detected one instance of vesivirus contamination of CRFK cells, using electron microscopy.

NOTE: We have also experienced a few cases where researchers working with cells that were contaminated with adventitious viruses had attributed cell deterioration to poor technique (graduate students were often blamed).

9. Engineered cell lines for virus isolation

As elegantly pointed out by Dr. Paul Olivo [68], rapid diagnostic assays based on direct detection of viral antigen or nucleic acid are used with increased frequency in clinical virology laboratories. Regardless, virus culture remains the only way to detect infectious virus and to analyze clinically relevant viral phenotypes, such as drug resistance. Growth of viruses in cell culture is costly, labor intensive and time-consuming and requires the use of many different cell lines. Transgenic technology offers the possibility of using genetically modified ("engineered") cell lines to improve virus growth in cell culture and to facilitate detection of virus-infected cells. Whereas various approaches are available, the two common applications of cell engineering for diagnostic virology are: (a) engineering of susceptible cell lines to over-express virus receptors, and (b) genetically modifying cells so that they express a reporter gene only after infection with a specific virus, allowing for the detection of infectious virus by rapid and simple enzyme assays such as β-galactosidase assays without the need for antibodies.

Conceptually, just because a cell line is susceptible and permissive for a virus does not also mean that the optimal virus receptor is present on the cell surface. The number of viral receptors on the cell surface might also be suboptimal. If an authentic gene for the viral receptor is over-expressed (through genetic engineering), virus attachment and entrance into the cell should be improved. For that reason, we have engineered Vero E6 and other cells (such as Mv1 Lu, which are mink lung cells), that over-express canine signaling lymphocyte activation molecule (cSLAM), thought to be the major virus receptor of *Canine distemper virus* (CDV). Essentially the same rationale was used by others in their decisions to engineer cSLAM-over-expressing Vero-derived cell lines [69, 70]. Analogous to the experience of the other groups, we find that many CDV strains are detected within one day of infection of the SLAM-expressing cells vs. up to one month in conventional Vero cells. A photograph of Mv1-cSLAM cells showing syncytia within 16 hrs of infection by an American type 2 CDV strain is shown in the figure below. A related cell line that expresses human SLAM, Vero-hSLAM, is used for the isolation of MeV [71], a virus closely related to CDV and is also often difficult to isolate using conventional Vero cells.

Similarly we and others have engineered MDCK cells that over-express a sialyl transferase 1 (SIAT1) that catalyze the formation of α-2,6 - linked sialic acid receptors recognized by human influenza A viruses. In our hands, most contemporary human influenza H1N1 and H3N2 viruses are detected earlier, and form higher virus titers in the engineered MDCK cells, than occurs in conventional MDCK cells (J. Lednicky and D. Wyatt, data to be presented elsewhere), in agreement with a recent report [72]. Apart from standard MDCK cells, we have also engineered additional cell lines that over-express SIAT1, and also, cell lines that over-express SIAT4 to catalyze the formation of α-2,3 - linked sialic acid receptors (data and information to be presented elsewhere). With these SIAT1 and SIAT4 over-expressing cells, we have isolated influenza viruses that others were unable to, and also, as for [72], we have recovered viruses from frozen stocks that others had great difficulty with.

Another example of a cell line that has been engineered for effective isolation of a virus is L20B, used for the isolation of poliovirus, which is an enterovirus. L20B cells are derived from mouse L cells, and were engineered to express the poliovirus receptor. One advantage of using these cells over human cell lines such as Hep2 and RD for the isolation of polioviruses is that few human enteroviruses can complete their replication cycle in mouse cells [73]. Thus, it is possible to engineer cells normally not permissive for a particular virus into a permissive version.

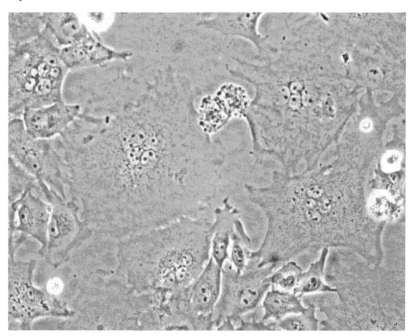

Figure 1. Mv1Lu-cSLAM cells 16 hrs post-inoculation with lung homogenate from a dog with distemper including neurologic signs. Multinucleated cells, a sign of CDV infection, are evident. An American type 2 CDV variant was isolated. Original magnification at 400X.

For the genetic modification of cultured cells (to express a virus receptor), it is no longer necessary to first clone a cDNA of a particular virus receptor in many instances. Many companies now sell full-length cDNAs of human (and non-human) genes already cloned into plasmid expression vectors. Some of the new generation expression vectors are also highly active across species, thus eliminating the need to optimize enhancer/promoter sequences in the expression vector.

The second general approach for the engineering of cells for virus detection/isolation, that of modifying cells to express reporter genes in response to virus infection, is still relatively new but has already proven useful for diagnostic virology. For example, the ELVIS HSV system marketed by the company Diagnostic Hybrids for the detection and typing of human herpes simplex viruses (HSV) 1 and 2. The ELVIS HSV system uses BHK cells engineered to express β-galactosidase in response to HSV infection. After the engineered cells are inoculated with clinical specimen, a blue precipitate forms over the infected cells due to interaction of β-galactosidase with a colorimetric substrate. Formation of a blue color facilitates detection of infected cells, with the net effect of reducing the time from sample inoculation to virus detection from (typically) seven or more days to one to two days. A subsequent immunofluorescence assay using monocolonal antibodies specific to HSV 1 or 2 is used to type the virus.

10. Paradigms for virus isolation

We are often asked "What is the best combination of indicator cells to use for the (primary) isolation of viruses from clinical specimens?" Of course, this is not a question that has one correct answer. Indeed, there is no magic combination that is universally applicable. And variability exists according to the availability of reagents in various countries. The best answer to the question of which cells to use is to rely on the experience of the virologists working on site, and to use as wide a variety of indicator cells as possible for the isolation of unknown viruses. There are however some guidelines that are applicable across many applications:

a. Many respiratory viruses replicate in the upper respiratory tract, and require temperatures lower than 37°C for optimal replication. For each type of indicator cell line (or primary cells), it is advisable to inoculate replicate cell cultures, and incubate one at 35° to 37°C, the other at 32° to 34°C.

b. Cultures should not be considered negative for virus isolation if CPE are not detected. A second measure should be considered and well-thought criteria should be developed before rejecting a "negative" culture. For example, would CPE form if the cultures were held for a longer period of time? Examples of secondary evaluations include performance of hemagglutination (using spent culture media) or hemadsorption tests (performed directly on the infected cells), and electron microscopy. Hemagglutination and hemadsorption tests are best performed using red blood cells from two unrelated species (e.g., one mammalian and one avian species). Electron microscopy should be performed using material from spent media to detect liberated virions, and also, on a sample of the infected cells (often, the number of liberated virions is too low to be easily visualized through electron microscopy, and virus infection is determined only by examining the infected cells themselves).

c. For all work, non-infected cultures (negative controls) should be maintained and examined in parallel with any virus-infected cultures.

d. Where economically feasible, new generation molecular tests should be used to assist in the identification of new viruses.

e. Some viruses require "adaptation" prior to adequate replication in cultured cells and the formation of CPE. In the past, the process referred to as "blind culture" was performed when virus was suspected but CPE inapparent. A popular version of this method is to periodically remove samples from a culture of presumably infected cells, and to inoculate that into a new batch of cells. This process is repeated four times. An adjunct to former process is to split the infected cells (if confluent) into a larger flask or into several flasks and allow the cells to replicate. This may make CPE apparent if actively replicating cells are optimal for the detection of the CPE caused by a particular virus.

f. The isolation of viruses from mosquitoes, ticks, etc., can be challenging. Many mosquito-borne viruses are best amplified to high titer in mosquito cell lines, where they often proliferate without causing CPE, prior to subculture in animal cells. Occasionally, it has also been necessary to inoculate newborn or suckling rodents (subcutaneously or intracerebrally) to obtain a high virus titer for the inoculation of cell cultures.

g. During the primary isolation of virus from clinical or environmental specimens, many laboratories routinely filter specimens though a 0.45μm filter prior to inoculation of cell cultures. This filtration step is performed to remove bacteria, fungi, and other potential microbial contaminants, and non-living particulates. A problem with this filtration step is that many viruses are pleomorphic and some have long, filamentous forms that may exceed 0.45 μm. This includes some influenza viruses, and morbilliviruses. Also, in clinical specimens, many viruses are attached to cellular and other debris, and are trapped by the filter. We recommend the inoculation of two batches of cells; one with a filtered aliquot, the other unfiltered, of the virus specimen. Unfortunately, it is not uncommon these days for bacteria in clinical specimens (such as normal flora that are contaminants of naopharyngeal swabs) to be resistant to penicillin and streptomycin; we prefer to use an antibiotic mixture that includes neomycin in addition to penicillin and streptomycin.

h. When economically possible, we suggest inoculation of cells growing on a relatively large growing surface with specimen. The idea behind this is to effect a dilution of the inoculum over a wide area, reducing the concentration of toxic agents. For example, add a small inoculum (e.g. 100 μl) to cells in a flask with a growing surface of 75 cubic centimeters (T75 flask). This is especially helpful when specimens must be pre-treated with a high concentration of antibiotics (such as when attempting the isolation of virus from a stool sample).

i.   A largely forgotten finding is that antibiotics can suppress the growth of some viruses. This concept is anti-dogmatic, but should always be considered when attempts at the isolation of a particular virus are not very successful. A recent example is described in [74].

j.   Some viruses remain cell-associated, as mentioned above, and for effective virus isolation, the virions must be liberated from the material they are attached to. This is one reason some clinical specimens designated for virus isolation are "vortexed" in the presence of glass beads prior to inoculation of the specimen onto cells. Instead of bead disruption, in some cases, freeze-thaw of the specimen can be used to dissociate the virions from the cells or debris they are attached to. Freeze-thaw cannot be used for herpesviruses and other viruses that are rapidly inactivated by such as process. Some viruses such as JC virus can be dissociated from cell debris by pretreatment of the specimen with neuraminidase. In any case, it is advisable to explore whether mechanical or biochemical options are available to improve virus isolation. However, for biosafety reasons, it is always best not to use procedures that create aerosols.

11. Closing remarks

In this article, we have discussed some of the art behind virus isolation through cell culture. Due to the complexities of cell culture, and the nature of the biomaterials used, it is not possible to consistently attain the same end results at all times. Moreover, viruses constantly mutate, and so the "rules of the game" can change. Therefore, the practice of cell culture for virus isolation is part art, part science, and part luck. Nevertheless, following principles mentioned in this manuscript, we have succeeded at maintaining cultured cells for long periods, and have isolated and propagated many "difficult" viruses.

Author details

John A. Lednicky*
Environmental and Global Health, University of Florida, Gainesville, Florida, USA

Diane E. Wyatt
KC Bio, LLC, E. Santa Fe, Olathe, Kansas, USA

12. References

[1] Schaeffer WI. Terminology Associated with Cell, Tissue, and Organ Culture, Molecular Biology, and Molecular Genetics. Tissue Culture Association Terminology Committee. In Vitro Cell. Dev. Biol. 1990; 26 (1):97-101.

[2] Hay RJ. Cell Line Availability: Where to Get the Cell Lines You Need. In: Mather JP, Barnes D (ed.), Animal Cell Culture Methods, Methods in Cell Biology, Volume 57, Academic Press, San Diego, California; 1998. p. 34.

* Corresponding Author

[3] Barallon R, Bauer SR, Butler J, *et al.* Recommendation of Short Tandem Repeat Profiling for Authenticating Human Cell Lines, Stem Cells, and Tissues. In Vitro Cell. Dev Biol. Anim. 2010; 46:727-732.

[4] Cabrera CM, Cobo F, Nieto A *et al.* Identity Tests: Determination of Cell Line Cross-Contamination. Cytotechnology 2006; 51:45–50.

[5] Barile MF, Kern J. Isolation of *Mycoplasma arginini* for Commercial Bovine Sera and its Implication in Contaminated Cell Cultures. Proc. Soc. Exp. Biol. 1971; 138(2):432-437.

[6] Chen TR. In Situ Detection of Mycoplasma Contamination in Cell Cultures by Fluorescent Hoeschst 33258 Stain. Exp. Cell. Res. 1977; 104:255 -262.

[7] Lincoln CK, Gabridge MG. Cell Culture Contamination: Sources, Consequences, Prevention, and Elimination. In: Mather JP, Barnes D (ed.), Animal Cell Culture Methods, Methods in Cell Biology, Volume 57, Academic Press, San Diego, California; 1998. pp. 49-65.

[8] Brunner D, Frank J, Appl H, *et al.* Serum-free Cell Culture: the Serum-free Media Interactive Online Database. *ALTEX.* 2010; 27:53-62.

[9] Erickson GA, Bolin SR, Landgraf JG. Viral Contamination of Fetal Bovine Serum Used for Tissue Culture: Risks and Concerns. Dev. Biol. Stand. 1991;75:173-175.

[10] Fong CK, Gross PA, Hsiung GD, Swack NS. Use of Electron Microscopy for Detection of Viral and Other Microbial Contaminants in Bovine Sera. J. Clin. Microbiol. 1975; 1(2):219-224.

[11] Jennings A. Detecting Viruses in Sera: Methods Used and Their Merits. Dev. Biol. Stand. 1999; 99:51-59.

[12] Kniazeff AJ, Wopschall LJ, Hopps HE, Morris CS. Detection of Bovine Viruses in Fetal Bovine Serum Used in Cell Culture. In Vitro 1975; 11(6):400-403.

[13] Rolleston WB. Bovine Serum: Reducing the Variables Through the Use of Donor Herds. Dev. Biol. Stand. 1999; 99:79-86.

[14] Wessman SJ, Levings RL. Benefits and Risks Due to Animal Serum Used in Cell Culture Production. Dev. Biol. Stand. 1999; 99:3-8.

[15] Bolin SR, Matthews PJ, Ridpath JF. Methods for Detection and Frequency of Contamination of Fetal Calf Serum with Bovine Viral Diarrhea Virus and Antibodies Against Bovine Viral Diarrhea Virus. J. Vet. Diagn. Invest. 1991; 3(3):199-203.

[16] Falcone E,Tollis M, Conti G. Bovine Viral Diarrhea Disease Associated with a Contaminated Vaccine. Vaccine. 1999; 18(5-6):387-8.

[17] Harasawa R, Mizusawa H. Demonstration and Genotyping of Pestivirus RNA from Mammalian Cell Lines. Microbiol. Immunol. 1995;39(12):979-985

[18] Levings RL, Wessman SJ. Bovine Viral Diarrhea Virus Contamination of Nutrient Serum, Cell Cultures and Viral Vaccines. Dev. Biol. Stand. 1991;75:177-181.

[19] Vilcek S. Identification of Pestiviruses Contaminating Cell Lines and Fetal Calf Sera. Acta. Virol. 2001; 45(2):81-86.

[20] Schuurman R, van Steenis B, Sol C. Bovine Polyomavirus, a Frequent Contaminant of Calf Serum. Biologicals 1991; 19(4): 265-270.

[21] van der Noordaa J, van Steenis B, van Strien A, *et al.* Bovine Polyomavirus, a Frequent Contaminant of Calf Sera. Dev. Biol. Stand. 1999; 99:45-47.

[22] Allander T, Emerson SU, Engle RE, *et al.* A Virus Discovery Method Incorporating DNase Treatment and its Application to the Identification of Two Bovine Parvovirus Species. Proc. Natl. Acad. Sci. USA 2001; 98, 11609–11614.

[23] Lau, SKP, Woo PCY, Tse H, *et al.* Identification of Novel Porcine and Bovine Parvoviruses Closely Related to Human Parvovirus 4. J. Gen. Virol. 2008; 89:1840–1848.

[24] Nettleton PF, Rweyemamu MM. The Association of Calf Serum with the Contamination of BHK21 Clone 13 Suspension Cells by a Parvovirus serologically related to the minute virus of mice (MVM). Arch. Virol. 1980; 64(4):359-74.

[25] Egyed L. Bovine Herpesvirus Type 4: A Special Herpesvirus. Acta. Vet. Hung. 2000; 48(4):501-13.

[26] Egyed L. Replication of Bovine Herpesvirus Type 4 in Human Cells In Vitro. J. Clin. Microbiol. 1998; 36(7):2109-2111.

[27] Fong CK, Landry ML. An Adventitious Viral Contaminant in Commercially Supplied A549 cells: Identification of Infectious Bovine Rhinotracheitis Virus and its Impact on Diagnosis of Infection in Clinical Specimens. J. Clin. Microbiol. 1992; 30(6):1611–1613.

[28] Michalski FJ, Dietz A, Hsiung GD. Growth Characteristics of Bovine Herpesvirus 1 (Infectious Bovine Rhinotracheitis) in Human Diploid Cell Strain WI-38. Proc. Soc. Exp. Biol. Med. 1976;151(2):407-410.

[29] Hansen G, Wilkinson R. Gamma radiation and virus inactivation: new findings and old theories. Art to Science 1993; 12(2):1–6.

[30] House C, House JA, Yedloutschnig RJ. Inactivation of Viral Agents in Bovine Serum by Gamma Irradiation. Can. J. Microbiol. 1990; 36:737–740.

[31] Wessman SJ, Levings RL. Collective experiences of adventitious viruses of animal-derived raw materials and what can be done about them. Cytotechnology 1998; 28:43–48.

[32] Wyatt DE, Keathley JD, Williams CM, Broce R. Is there life after irradiation? Part 1: Inactivation of Biological Contaminants. BioPharm, 1993; June, pp. 34-40.

[33] Wyatt DE, Keathley JD, Williams CM, *et al.* Is there life after irradiation? Part 2: Gamma-irradiated FBS in Cell Culture." BioPharm, 1993; July-August, pp. 46 – 52.

[34] Gauvin G, Nims R. Gamma-Irradiation of Serum for the Inactivation of Adventitious Contaminants. PDA J. Pharm. Sci. Technol. September/October 2010; 64:432-435.

[35] Nims RW, Gauvin G, Plavsic M. Biologicals. Gamma Irradiation of Animal Sera for Inactivation of Viruses and Mollicutes--a Review. 2011; 39(6):370-377.

[36] Gould MC. Endotoxin in Vertebrate Cell Cultures: Its Measurement and Significance. In: Uses and Standardization of Vertebrate Cell Cultures, *In Vitro* Monograph number 5, 1984, Tissue Culture Association, Gaithersburg, MD: 125 136.

[37] Price PJ, Evege EK. Serum-free Medium Without Animal Components for Virus Production. Focus, 19 (1997), pp. 67–69.

[38] Yuk IH, Lin GB, Ju H, *et al.* A Serum-free Vero Production Platform for a Chimeric Virus Vaccine Candidate. Cytotechnology 2006; 51:183-192.

[39] Butler M, Burgener A, Patrick M, *et al.* Application of a Serum-free Medium for the Growth of Vero Cells and the Production of Reovirus. Biotechnol. Progress 2000; 16:854–858.

[40] Hu AY-C, Tseng Y-F, Weng T-C, *et al.* 2011. Production of Inactivated Influenza H5N1 Vaccines from MDCK Cells in Serum-free Medium. PLoS One 2011; 6:e14578.

[41] Frazzati-Gallina NM, Paoli RL, Mourão-Fuches RM, *et al.* Higher Production of Rabies Virus in Serum-free Medium Cell Cultures on Microcarriers. J Biotechnol. 2001; 92(1):67-72.

[42] Armstrong SE, Mariano JA, Lundin DJ. The Scope of Mycoplasma Contamination Within the Biopharmaceutical Industry. Biologicals 2010, 38:211-213.

[43] Polak-Vogelzang AA, Angulo AF, Brugman J, Reijgers R. Survival of Mycoplasma hyorhinis in Trypsin Solutions. Biologicals 1990, 18:97-101.

[44] Rottem S, Barile MF. Beware of Mycoplasmas. Trends Biotechnol 1993, 11:143-151.

[45] Barile MF, Razin, S. Mycoplasmas in Cell Culture. In Rapid Diagnosis of Mycoplasmas (Kahane I., and Adoni, A., eds) 1993 pp. 155-193, Plenum Press, New York.

[46] Windsor H IRPCM report: Prevention and Control of Mycoplasma Contamination in Cell Cultures. 2010. http://www.the-iom.org/assets/files/IRPCM_Team_Mycoplasma _contamination.pdf

[47] Windsor HM, Windsor GD, Noordergraaf JH. The Growth and Long Term Survival of Acholeplasma laidlawii in Media Products Used in Biopharmaceutical Manufacturing. Biologicals 2010, 38:204-210.

[48] McGarrity GJ, Kotani H, Butler GH. Mycoplasmas and Tissue Culture Cells. In: Maniloff J, McElhaney RN, Finch LR, Baseman JB (eds). Mycoplasmas: Molecular Biology and Pathogenesis. American Society for Microbiology, Washington, DC; 1992. pp. 445-454.

[49] Razin S, Yogev D, Naot Y. Molecular Biology and Pathogenicity of Mycoplasmas. Microbiol. Mol. Biol. Rev. 1998; 62:1094-1156.

[50] Namiki K, Goodison S, Porvasnik S, *et al.* Persistent Exposure to Mycoplasma Induces Malignant Transformation of Human Prostate Cells. PLoS One 2009, 4:e6872.

[51] Drexler HG, Gignac SM, Hu ZB, *et al.* Treatment of Mycoplasma Contamination in a Large Panel of Cell Cultures. In Vitro Cell Dev Biol Anim. 1994, 30A:344 347.

[52] Somasundaram C, Nicklas W, Matzku S. Use of Ciprofloxacin and BM-Cyclin in Mycoplasma Decontamination. In Vitro Cell. Dev. Biol. 1992, 28A:708-710.

[53] Uphoff CC, Drexler HG. Eradication of Mycoplasma Contaminations. Methods Mol Biol. 2005, 290:25-34.

[54] Timenetsky J, Santos LM, Buzinhani M, Mettifogo E. Detection of Multiple Mycoplasma Infection in Cell Cultures by PCR. Braz. J. Med. Biol. Res. 2006, 39(7):907-14.

[55] Del Giudice RA, Gardella RS. Antibiotic Treatment of Mycoplasma Infected Cell Cultures. In: Tully JG, Razin S (Ed), Molecular and dagnostic procedures in mycoplasmology. Academic Press, San Diego; 1996, pp. 439-443.

[56] Hotta H, Nihei K, Abe Y, et al. Full-length Sequence Analysis of Subacute Sclerosing Panencephalitis (SSPE) Virus, a Mutant of Measles Virus, Isolated from Brain Tissues of a Patient Shortly after Onset of SSPE. Microbiol. Immunol. 2006; 50(7):525-34.

[57] Hara K, Sugimoto C, Kitamura T, *et al.* Archetype JC Virus Efficiently Replicates in COS-7 Cells, Simian Cells Constitutively Expressing Simian Virus 40 T Antigen. J. Virol. 1998, 72(7):5335-42.

[58] Abiko C, Mizuta K, Itagaki T, *et al.* Outbreak of Human Metapneumovirus Detected by Use of the Vero E6 Cell Line in Isolates Collected in Yamagata, Japan, in 2004 and 2005. J. Clin. Microbiol. 2007, 45:1912-1919.

[59] Marcus-Sekura C, Richardson JC, Harston RK, *et al.* Evaluation of the Human Host Range of Bovine and Porcine Viruses that May Contaminate Bovine Serum and Porcine Trypsin Used in the Manufacture of Biological Products. Biologicals 2011, 39(6):359-69.

[60] Teixeira TF, Dezen D, Cibulski SP, *et al. Torque teno sus virus* (TTSuV) in Cell Cultures and Trypsin. PLoS One 2011, 6(3):e17501.

[61] Hattermann K, Roedner C, Schmitt C, *et al.* Infection Studies on Human Cell Lines with Porcine Circovirus Type 1 and Porcine Circovirus Type 2. Xenotransplantation 2004 , 11(3):284-294.

[62] Ma H, Shaheduzzaman S, Willliams DK, *et al.* Investigations of Porcine Circovirus Type 1 (PCV1) in Vaccine-related and Other Cell Lines. Vaccine 2011, 29(46):8429-8437.

[63] Tischer I, Bode L, Apodaca J, et al. Presence of Antibodies Reacting with Porcine Circovirus in Sera of Humans, Mice, and Cattle. Arch Virol. 1995, 140(8):1427-1439.

[64] Nims RW. Detection of Adventitious Viruses in Biologicals – A Rare Occurrence. Dev. Biol. (Basel) 2006; 123: 153-164; discussion 183-197.

[65] Gillet L, Minner F, Detry B, *et al.* Investigation of the Susceptibility of Human Cell Lines to Bovine Herpesvirus 4 Infection: Demonstration that Human Cells Can Support a Nonpermissive Persistent Infection Which Protects Them Against Tumor Necrosis Factor Alpha-Induced Apoptosis. J. Virol. 2004, 78(5):2336-2347.

[66] Yamane D, Zahoor MA, Mohamed YM, et al. Activation of Extracellular Signal-regulated Kinase in MDBK Cells Infected with Bovine Viral Diarrhea Virus. Arch Virol. 2009; 154(9):1499-1503.

[67] Berting A, Farcet MR, Kreil TR. Virus Susceptibility of Chinese Hamster Ovary (CHO) Cells and Detection of Viral Contaminations by Adventitious Agent Testing. Biotechnology and Bioengineering 2010, 106 (4) 598–607.

[68] Olivo, P. D. Transgenic Cell Lines for Detection of Animal Viruses. Clin. Microbiol. Rev. 1996, 9:321-334.

[69] Seki F, Ono N, Yamaguchi R, Yanagi Y. Efficient Isolation of Wild Strains of Canine Distemper Virus in Vero Cells Expressing Canine SLAM (CD150) and Their Adaptability to Marmoset B95a cells. J. Virol. 2003, 77:9943-9950.

[70] von Messling V, Springfeld C, Devaux P, Cattaneo R. A Ferret Model of Canine Distemper Virus Virulence and Immunosuppression. J. Virol. 2003, 77:12579-12591.

[71] Ono N, Tatsuo H, Hidaka Y, *et al.* Measles Viruses on Throat Swabs from Measles Patients Use Signaling Lymphocytic Activation Molecule (CDw150) but Not CD46 as a Cellular Receptor J. Virol. 2001, 75:4399-4401.

[72] Oh, DY, Barr IG, Mosse JA, Laurie KL. MDCK-SIAT1 Cells Show Improved Isolation Rates for Recent Human Influenza Viruses Compared to Conventional MDCK Cells. J. Clin. Microbiol. 2008, 46:2189-2194.

[73] Wood DJ, Hull B. L20B cells simplify culture of polioviruses from clinical samples. J. Med. Virol. 1999, 58(2):188-192.

[74] Asada M, Yoshida M, Suzuki T, *et al*. Macrolide Antibiotics Inhibit Respiratory Syncytial Virus Infection in Human Airway Epithelial Cells. Antiviral Res. 2009, 83(2):191-200.

Chapter 10

Tissue Culture to Assess Bacterial Enteropathogenicity

Aurora Longa Briceño, Zulma Peña Contreras, Delsy Dávila Vera, Rosa Mendoza Briceño and Ernesto Palacios Prü

Additional information is available at the end of the chapter

1. Introduction

It is quite clear nowadays that the pathogenesis of infectious disease, although determined by pathogenic features of their causative agents, cannot be fully comprehended without structural (morphologic) analysis of immediate interaction between these agents and cells, tissues, and defense systems of the host. In diseases caused by agents adapted to certain cellular targets, the pathogenesis is, in addition, influenced by histophysiologic features of these targets wich are utilized or distorted or both by the pathogen.

Acute diarrheal diseases caused by Gram-negative bacteria: *Vibrio cholerae*, diarrheagenic *Escherichia coli*, *Shigella*, *Salmonella*, *Aeromonas* and enteropathogenic *Yersinia*, are examples of this type of infection. All of these agents target the same cell type, the enterocyte, and produce potent exotoxinas. [1]

The morphologic method either produced an impetus to bacteriologic studies or served as a tool in the evaluation of bacterial pathogenicity, including that of genetically altered microorganisms. In particular, morphologic studies have revealed that enteric bacteria either colonize enterocytes while remaining epicellular (i.e., bound to the cell surface) with or whitout affecting cellular architecture, or they invade the cell with or whitout its destruction. On the other hand, biochemical studies have shown that prevalence of the secretory or destructive inflammatory disturbances in the gut is, for the most part, determined by bacterial exotoxins. Although exotoxins affect enterocytes by a variety of mechanisms, they can be combined into two groups that are referred to as cytotonic and cytotoxic or respectively, as enterotoxins and cytotoxins. It has also been established that each step in host-pathogen interaction is governed by multiple determinants encoded in bacterial plasmid and chromosome genes. [1,2,3]

The genus *Aeromonas* comprises gram-negative bacteria that can be isolated from water and a diversity of foods. Some strains are important diarrhea producers, particularly in children under five years and in older adults [4,5].

The clinical manifestations of diarrhea vary from autolimited symptoms to severe cases with presence of mucus and blood in faeces, suggesting that, as in *Escherichia coli* pathogenic types, *Aeromonas* virulence is multifactorial [6,7]. In recent years *Aeromonas* spp. have emerged as an important human pathogen, with increasing incidence among travelers (causative agent of traveler's diarrhea), due to their presence in food as well as in treated water for human consumption [8].

The aim of this chapter, is to show that with the development of an *in vitro* animal model, to explore the mechanisms related to the colonization of the digestive tract, as well as the determination of the mechanisms of interaction with the host epithelium, provide a valuable tool in the study of bacterial entropathogenicity that the genus *Aeromonas*

2. Interaction host/bacterial pathogen

The *Aeromonas* enteropathogenic capability is usually underrated [7,9,10]. Information about decisive diarrhea virulence factors is limited, due to lack of appropriate animal models for the study of *Aeromonas* genus. [7,10,11]). With the models developed up to now, it has obtained scarce information about the interaction between the host and *Aeromonas* spp. [10,11,12,13,14].

In this regard, the development of an *in vitro* model to explore the mechanisms related to the colonization of the digestive tract by *Aeromonas* spp. as well as the determination of the mechanisms of interaction with the host epithelium, provides a valuable tool in the study of *Aeromonas* pathogenicity. On the other hand, the critical steps in the pathogenesis of virulent strains involve the adhesion and colonization of the intestinal mucosa by *Aeromonas*.

The present focuses in the analysis of the mechanisms causing the pathogenicity of *Aeromonas* strains in co-cultures of isolated bacteria with intestinal cells. Ultrastructural aspects of mouse small intestinal tissue cultures infected with *Aeromonas* strains are described using a novel experimental procedure which allows to culture *Aeromonas* inside a previously sterilized short cylinder of mouse's small intestine.

3. Materials and methods

3.1. Bacterial strains and growth conditions

Two strains of *Aeromonas caviae* were used, one (A) isolated from an asymptomatic patient and the other one (D) from a patient with diarrhea in which the isolated *Aeromonas* was the only enteropathogen. The strains were inoculated in trypticase soya agar (HIMEDIA laboratories Ltd., Bombay, India 400 086) at a concentration of 1.5 X 10^8 CFU/ml and incubated for 24h. After incubation the media were gently filtered and the corresponding *Aeromonas* strain resuspended in 1ml Basal Medium Eagle (BME).

3.2. Tissue culture procedure

Segments of the small intestine were removed from the abdominal cavity of young adult NMRI mice to prepare small intestine cylinders of ~3cm in length that were sterilized with a 10% chlorine solution. One end of each intestinal cylinder was tied with sterile surgical thread and the cavity was filled with 1.5 X 10^8 CFU/ml of a given one of the two *Aeromonas* strains suspended in BME. After filling, the cylinders were tied close with surgical thread. The whole preparations were co-cultured in 50ml of tissue culture media containing 90% BME with Earle´s salts and L-glutamine, 10% horse serum, 5000IU/ml penicillin and 5mg/ml streptomycin, 360mOsm, pH 7.2, at 37°C, under constant rotation at 70rpm, for 24h and 48h. During the incubation period, the media were oxygenated every 4h during daytime. Control small intestine cylinders were cultured for 48h under the same conditions but without *Aeromonas* inside. Once the programmed culture time was accomplished, the intestinal cylinders were removed from the flask and immediately immersed in a fixing solution containing 3% glutaraldehyde and 3% formaldehyde in 0.1M cacodylate buffer, pH 6.3 [15] during 6h at 4°C.

The intestinal cylinders were cut in small sections of approximately $3mm^3$, washed in 0.1M cacodylate buffer, pH 7.2, and postfixed for 24h in 1% osmium tetroxide prepared in the same buffer. The tissue was then dehydrated in ascending concentration ethanol solutions, followed by propylene oxide and finally embedded in Epon 812. Sections of 1μm were stained with 1% toluidin blue and observed under a high resolution light microscope. Sections of 90 nm were contrasted with uranyl acetate [16] and lead citrate [17] using a modification of this classic method [18] and were analyzed using a Hitachi-7000 transmission electron microscope. [19,20].

4. Effects of *Aeromonas caviae* co-cultured in mouse small intestine

The experimental desing herein presented offers appropriate conditions for the physiopathologic study of the interrelationship between the intestinal wall and bacteria under *in vitro* conditions very similar to *in situ* conditions.

On the other hand, this experimental model employed using co-cultures of mouse intestinal mucosa with *Aeromonas caviae* revealed important information on the pathogenicity of this species in gastrointestinal infections. The strain A, isolated from asymptomatic patient, produce minor to mild alterations of the intestinal wall, while more pronounced alterations were found at both periods of incubations when strain D, from the patient with diarrhea, was used to prepare the co-cultures.

The damage produced was demonstrated by the alterations of the integrity of intestinal microvilli, disruption of the epithelium and presence of mucosal microulcerations, which were brought about by the fact that this strain is an high cytotoxin producer and may possess other virulence factors [4,21]

These strains belong to the *A. caviae* species and were considered of lesser virulence than other strains of this species [22].

Light microscopical observations revealed a varying degree of histological alterations of the intestinal wall, according to the severity of the damage. The major tissue damage is shown in Figure 1, where a large bacterial cluster can be seen occupying the crypts between intestinal folds. In most cases, the intestinal wall showed a high degree of generalized cellular atrophy with tissue lysis when the diarrhea producing strain (D) was used. These images were seen at 24h as well as at the 48h samples of incubation.

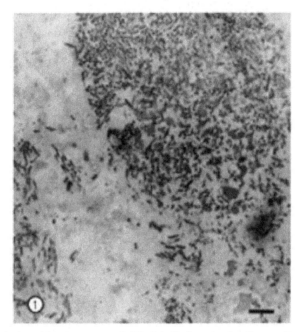

Figure 1. Large clusters of Aeromonas occupied the intestinal crypts when the co-cultures were prepared using the strain D. which comes from symptomatic patient. Note: the high degree of tissular lysis. Bar: 9.4μm.

In cases of moderate damage, the basic cytoarchitecture of the intestinal mucosa was preserved and it was possible to recognize enterocytes, mucous cells and germinative or mother cells. The villi as well as the majority of the crypts were seen, although some of them were shorter and thicker (Figures 2 and 3).

These mild damages were caused when the strain A from the asymptomatic species was employed for the co-culture. Microvilli were seen forming part of the apical surface of the enterocyte as well as cytoplasmatic protrusion that come from the apical portion of the enterocytes (Figure 4).

Some segments of the intestinal cylinder co-cultured with *Aeromonas* strains showing a higher degree of alterations contains clusters of spheroidal cells associated to bacterial elements (Figures 3 and 6).

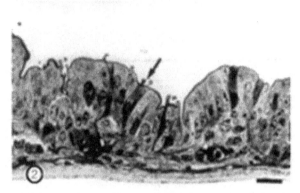

Figure 2. A case of mild damage of the small intestinal epithelium produced when the asymptomatic strain (A) was co-cultured during 48h. Well-aligned enterocytes can be seen, with a few of them exhibiting a dark content indicating cellular atrophy. The arrow points at an exocytic fragment. Bar 24.0μm

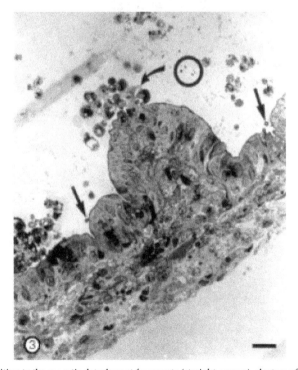

Figure 3. In addition to the exocytic detachment fragments (straight arrows), clusters of extraintestinal cells close to the epithelium are also seen (curved arrow). Some *Aeromonas* (circle) are present. Note that bacterial cells are not in contact with the entenocytes. Strain A was used co-cultured. Bar: 24.0um.

Transmission electron microscopy revealed minor alterations of the intestinal mucosa when the strain from asymptomatic patients (A) was co-cultured. Most enterocytes were seen with seen with typical ultrastructural characteristics, however, they showed numerous apical protrusion detachments (Figures 2 and 4). In more damaged regions, there was a progressive atrophy of the epithelial cells showing loss of microvilli and large cellular vacuoles loaded with cellular detritus (Figure 5).

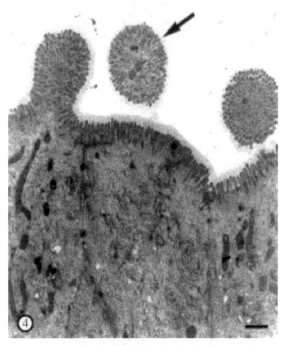

Figure 4. Enterocytes show almost normal ultrastmctural characteristics at 24h of co-culture when the asymptomatic strain (A) was employed. The main pathological feature observed in these cultures is the presence of exocytic detachments (arrow). The enterocytic epithelial cell, cytoplasm and microvilli also have normal ultrastructural features. Bar: 1.0μm.

In the intestinal regions having intermediate epithelial alterations, globular cells identified as blood and lymphatic cells were observed in the gut lumen associated to *Aeromonas* and cellular debris (Figures 3 and 6). Eosinophils were seen with multilobular nuclei and their characteristic lysosomes with crystal-like structures (Figures 7 and 8). In Figure 8 phagocytated *Aeromonas* within the eosinophilic cytoplasm are clearly visible.

Cells identified as lymphocytes (Figure 7) and others as plasma cells (Figure 9) were also observed as part of the clusters of globular cells found in the small intestine cylinder lumen after two days of culture.

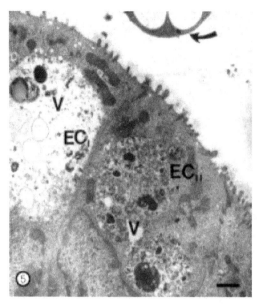

Figure 5. After 48h, the co-cultures prepared with strain A, the enterocytes showed large vacuoles (V) loaded with cytoplasmic debris ('EC,: EC") as seen in this photograph. Note the significant reduction of the microvilli. Bar: 1.0pm.

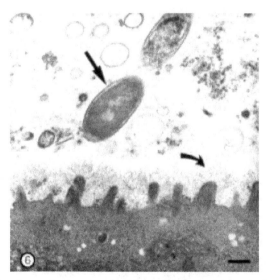

Figure 6. In all analyzed cases. the *Aeromonas* (straight arrow) are not seen attached to the surface of enterocytes that. while showing show a significant reduction of microvilli, have preserved their glycocalix cover (curved arrow). Bar: 0.4μm.

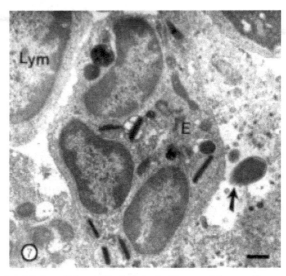

Figure 7. Among the globular cells located in the intestinal lumen of the co-cultures prepared using strain A from the asymptomatic patient. abundant eosinophilic granulocytes (B) were found. Numerous cytoplasmic lysosomes containing crystal-like inclusions characterize these cells and nuclear lobules are also visible. Profiles of *Aeromonas* (arrow) are detected outside the eosinophilic cells. Lym: lymphocyte. Bar: 0.5 μm.

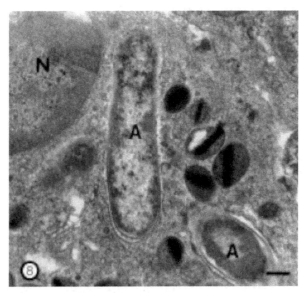

Figure 8. Within the cytoplasm of eosinophils. *Aeromonas* ('A') could be observed in association with the lysosomes. indicating the phagocytic capability of these cells. N. nucleus. Bar: 0.7 in.

In Figure 10 an image is shown of a cylinder of small intestinal segment incubated for 48h using the same culture conditions but without *Aeromonas* in its lumen, with the purpose to compare the control cylinders to the ones co-cultured with the bacteriae. No atrophic cells are seen and there are no visible epithelial protrusions, extraepithelial cells nor microulcerations of the intestinal wall. The villi and crypts show minor changes due to the culture conditions.

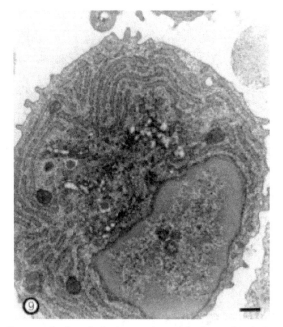

Figure 9. Plasma cells were other lymphatic elements frequently observed in the lumen of clustered cells. In this particular case. note the proliferation of the rough endoplasmic reticulum. Bar: 0.4 μm.

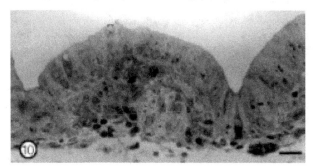

Figure 10. A control small intestine cylinder incubated without *Aeromonas* during 48h. Note the well preserved villi and cryptae as well as the absence of atrophic enterocytes and epithelial protrusions. Extraepithelial cells are not seen. Bar: 24.0 μm.

The cylinders of small intestinal cultures incubated for 24h and 48h revealed good preservation. The glycocalyx was clearly observed covering the microvilli (Figure 11). No vesicular chains were observed in or between the microvilli, nor was any type of bacteria observed in the intestinal lumen.

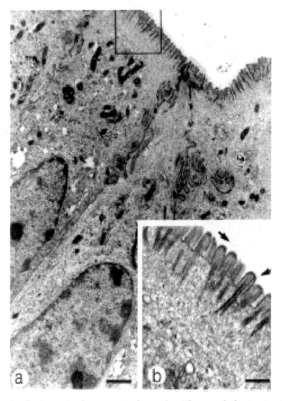

Figure 11. a: Segments of enterocytes from a control sample with normal ultrstructural characteristics alter 48h of culture. Bar: 1.5μm. b: detail of the normal ultrastructural features of an intestinal epithelial cell. Note the absence of vesicles between microvilli. Arrows, glycocalyx. Bar: 0.5 μm

After 24h de incubation, the *Aeromonas* spp. strain cultured with the intestinal tissue showed vesicular elements attached to the bacterial outer membrane (Figure 12). These vesicles were seen alone, in pairs, or in groups of five or more vesicular units organized in chains emerging from the external membrane. Adjacent to these vesicles some complex membranous structures were observed (Figure 12).

The small intestine tissue cultivated with *Aeromonas* spp. and incubated for 24h did not present any alteration or modification in its cellular structure. The only outstanding observation was the presence of vesicular chains composed of four to ten vesicular units, which were aligned between microvilli spaces (Figure 13).

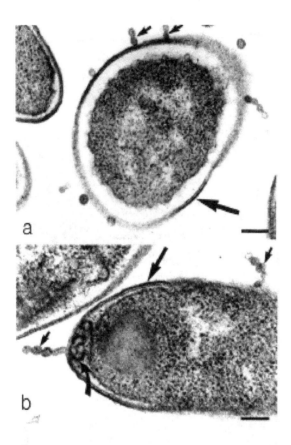

Figure 12. Cross sections of *Aeromonas* spp. cultured for 24h. Short chains of vesicles (in a) and large chains of vesicles (in b) can be seen emerging from the external bacterial membrane. Notice adjacent complex membranous structures (curved arrow) to these chains. Thick arrows: bacterial external membrane; short arrows: chains of vesicles. Bars: 0.17 µm in a, 0.15 µm in b.

When the segments of the small intestine were cultivated with *Aeromonas* spp., and then incubated for 48h, severe damage of the enterocytic epithelial surface was seen, with regions of tissular lysis on the surface of the intestinal microvilli (Figures 14 and 15). Numerous defensive cells such as lymphocytes were also observed (Figure 15), as well as abundant eosinophils showing their typical crystal-like structures inside lysosomes (Figure 16), and phagocytic mononuclear cells (Figure 17). Inside the cytoplasm of both of these cell types, *Aeromonas* spp. could be observed (Figures 16 and 17)

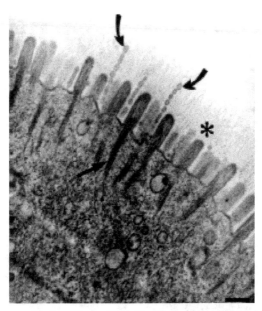

Figure 13. At a higher magnification, it is possible to identify the glycocalyx(asterisk) of the microvilli, and the particular vesicular chains (curved arrows) appearing between microvilli after 24h of culture with an Aeromonas spp. strain. Some microvilli fibrillary roots (straight arrow) are clearly seen. Bar: 0.35μm

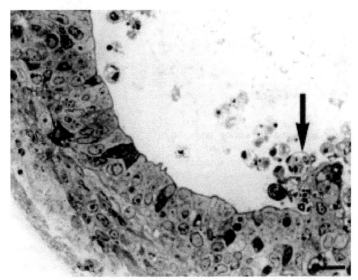

Figure 14. Light microscopic image of a segment of a small intestinal cylinder cultivated with Aeromonas spp. for 48h. The enterocytic epithelial surface shows regions with tissular lysis (arrow). Bar: 7.0 μm.

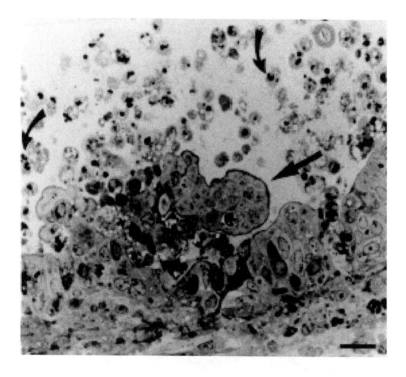

Figure 15. Notice the lysis (straight arrow) of the surface of the small intestine and the presence of defensive cells or lymphocytes (curved arrows) after 48h culture with *Aeromonas* from the symptomatic patient. Bar: 4.4 μm.

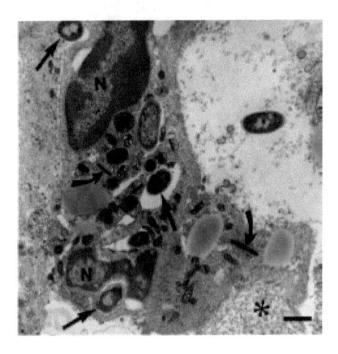

Figure 16. Eosinophils containing numerous phagocyted *Aeromonas* spp. (straight arrows) from symptomatic strain, are a common feature of the 48h intestinal *Aeromonas* cultures. These eosinophils show lysosome crystal-like structures (curved arrows). N: nucleus, *: cellular debris. Bar: 0.95 μm.

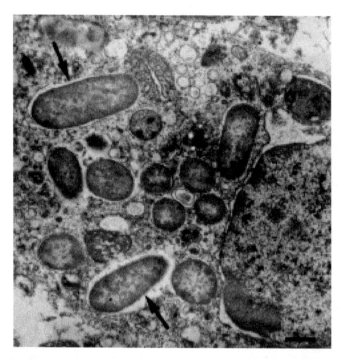

Figure 17. After 48h culture of *Aeromonas* spp. with small intestine tissue it is possible to see, within cellular clusters adjacent to the enterocytes, numerous phagocyte mononuclear cells. Inside the cytoplasm of these cells, phagocyted *Aeromonas* (arrows) can be identified. Bar:0.65 µm.

The pathogenicity of *Aeromonas* has been confirmed with this procedure and its virulence has been corroborated as strain-specific and not species-specific.

When the intestinal epithelium lesions were observed no bacteriae adhered to the cells were seen, indicating that direct bacteria-epithelial cell contact was nor required for tissue alterations: on the contrary, it seems that *Aeromonas* triggers a chemotactic response which activates migratory actions to the lymphatic cells from the adjacent lymphatic plaques.

The lymphatic submucous plaques and the autonomous defensive structures are able to act in culture conditions even in the absence of the circulatory blood elements, bone marrow cells or other lymphatic organs. Among the defense elements that migrate in answer to the chemotactic stimulus are the eosinophilic granulocytes, which have been reported to participate only in parasitic diseases. In this investigation their capability to defend intestinal cells against bacterial aggressions [23,24] is ultrastructurally documented.

Under the conditions of the experiment, no alteration of the intestinal epithelium was observed in the cultures from the small intestine with *Aeromonas* spp. incubated during 24h, except for vesicular chains aligned between microvilli spaces (Figure 13), which were not distinguished from those vesicular elements seen attached to the bacterial outer membrane (Figure 12) in the same cultures.

On the other hand, in cultures with 48h incubation, *Aeromonas* initially induced important tissue damage, including necrosis and lysis of the intestinal microvilli (Figures 14 y 15).

The vesicles produced on the surface of the bacterial outer membrane are also constituted by a double membrane unit. The production of these vesicles was observed in all the cultures analyzed. However, the more important vesicular formation was detected in the cultures incubated for 24h, whereas in the samples incubated for 48h, the vesicles were found where the intestinal damage was more severe.

The ultrastructural analysis suggest that *Aeromonas* spp. produce vesicles that form chains and can be seen between the microvilli of the intestinal epithelium. The contact between them could be one the mechanisms that trigger virulence factors produced by *Aeromonas* spp., contained in the vesicles, and that could subsequently initiate mechanisms that attract eosinophils and mononuclear cells into the intestinal lumen.

The vesicles do not appear to be intestinal exocytic vesicles because they are formed on the surface of the bacterial outer membrane. Moreover, no clathrin-like outer skeleton is seen, which suggest that this vesicular system is a part of the pathogenic mechanism of action that induces the migration of eosinophils and macrophages.

In the cultures with a 24h incubation period, the vesicular chains were found adhered to the bacterial outer membrane as well as occupying the free spaces between microvilli. In cultures incubated for 48h, the vesicles were absent, whereas some regions of the enterocytic epithelial surface lost their organization, leading to tissue lysis. The vesicles could constitute a carrier tool that facilitates the approximation and the interchange of enterotoxins, inmunologicals material or any other natural element responsible for the virulence and pathogenicity of *Aeromonas* spp., similar to what has been described by [25].

The strain isolated from the patient with diarrhea produced important alterations in the intestinal mucosa, indicating its enteropathogenic potential. This strain is a cytotoxin producer and, as has been pointed out for *Shigella,* a correlation exists between the severity of the lesions of the mucosal membrane and the toxigenicity of a particular strain [1].

5. Conclusion

In conclusion the genus *Aeromonas* has great enteropathogenic potential, although additional studies are necessary to understand more about the pathogenicity of other *Aeromonas* strains, this procedure also may useful to determinate currently unknown interactions between a variety of other enteropathogenic bacteria and, the results obtained in these studies give a valuable contribution for the future application of this procedure in the bacterial patogenicity research.

Chapter 11

Tissue-Based Model of HCV Replication as a Replacement for Animal Models in Drug Testing

Paulina Godzik

Additional information is available at the end of the chapter

1. Introduction

It is estimated that hepatitis C virus (HCV) has infected at least 170 million people worldwide [1]. More than 80% of infected patients develop the chronic infection, which leads to serious liver diseases, such as liver cirrhosis and hepatocellular carcinoma [2,3]. Chronic HCV infections may proceed with asymptomatic or with non-specific symptoms (fatigue, depression) for many years. Despite increasing knowledge about the virus pathogenesis, there is still no vaccine and successful antiviral therapy. The current standard of chronic hepatitis C therapy based on pegylated IFN-α 2a (PegIFN-α) and ribavirin (RBV) has limited efficacy and undesirable side effects. There is an urgent need for more effective therapies.

HCV is classified as a member of the Flaviviridae family and serves as a sole member of genus Hepacivirus. The genome of this enveloped virus consists of a single-stranded positive-sense RNA of approximately 9.6kb, which contains an open reading frame (ORF) encoding a polyprotein precursor of around 3000 amino acids (Figure 1). This single, large ORF is flanked by well conserved 5′ and 3′ untranslated regions (UTRs) [4]. The HCV 5′ UTR is a highly structured element, which includes internal ribosome entry site (IRES), a fragment required for genome translation [5,6]. The 3′ UTR contains short variable region, poly(U/UC) tract and X-tail region. The X-tail region forms highly conserved three stable stem-loop structures, which together with the poly(U/UC) are crucial for RNA replication [7,8]. Besides viral UTRs which are significant for translation and RNA replication, rest of the genome is diverse among several HCV isolates. According to genome differences, HCV isolates are divided into six particular genotypes, that differ in their nucleotide sequences by 31-34% [9]. Infections with genotype 1 are the most prevalent and dangerous, leading to liver injury and hepatocellular carcinoma [10-12]. Patients infected with genotypes 1, 4, 5 and 6 respond to treatment less effectively than patients infected with genotypes 2 and 3 [12,13].

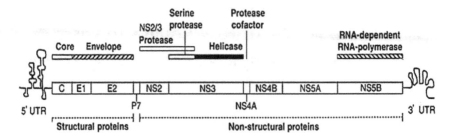

Figure 1. Genome organization of HCV

The HCV polyprotein precursor is co- and post-translationally processed into 10 proteins, that are divided into two groups, structural proteins: core protein, two glycoproteins (E1, E2), p7; and non-structural proteins: NS2, NS3, NS4A, NS4B, NS5A and NS5B. The HCV core protein forms viral nucleocapsid and also regulates cellular and viral gene expression, cell transformation and apoptosis by the interaction with cellular proteins and signaling pathways [14,15]. The two glycoproteins, E1 and E2, are essential components of the virion envelope and necessary for viral entry [16]. E2 plays a crucial role in attachment of the virus to a cell, by the interaction with host receptor – CD81 [17,18]. The p7 protein is a small polypeptide that forms an ion channel, suggesting that it belongs to the viroporin family [19,20]. The NS2 protein is a transmembrane protein, which together with amino-terminal domain of the NS3 protein create NS2-3 protease that cleaves the site between NS2 and NS3 during processing of the polyprotein [21]. The NS3 and NS4A (NS3-NS4A) built a multi-functional complex essential for viral polyprotein processing and RNA replication. The NS4A protein is a cofactor of NS3 protease activity and catalyzes polyprotein cleavage. The 442 C-terminal amino acids of the NS3 is a helicase that plays a crucial role in RNA replication [22]. The NS4B is an integral membrane protein that serves as a membrane anchor for the replication complex [23]. The NS5A protein is a membrane-anchored phosphoprotein, important in viral replication, but exact function of this protein is not known [22]. The last protein, NS5B is a viral RNA-dependent RNA polymerase which is crucial for RNA replication.

The first step of the HCV life cycle is attachment of the virus to a cell via E2 and cell receptor interaction [Figure 2]. CD81 has been the most extensively studied as a putative HCV receptor [17]. Several cell surface molecules, like: scavenger receptor B type I (SR-BI), low-density lipoprotein receptor (LDL-R) or asialoglycoprotein receptor (ASGP-R) have been also proposed to mediate HCV binding [24-27]. HCV entry into cells by pH-dependent endocytosis, but the mechanism of HCV fusion remains controversial [27,28]. Released viral RNA into the cytoplasm of infected cell serves directly as messenger RNA in an internal ribosome entry site-directed translation of the HCV polyprotein. The precursor polyprotein, which is targeted to the ER membrane is processed by cellular and viral proteases into 10 proteins. HCV replication starts with synthesis of complemantary negative-strand RNA which then serves as a template for production of numerous positive-strand genomic RNAs, both steps are catalyzed by the NS5B RNA-dependent RNA polymerase (RdRp).These RNAs serve as mRNA in translation, as a template for synthesis more negative strands and as substrates for viral assembly. Little is known about HCV assembly and release. It is suggested that core protein is sufficient for viral assembly and its interaction with RNA may play a role in switching from RNA replication to packaging [29,30]. After viral RNA association with core, viral nucleocapsids are formed and bud into the ER. Newly produced virions may leave the cell through the constitutive secretory pathway [28].

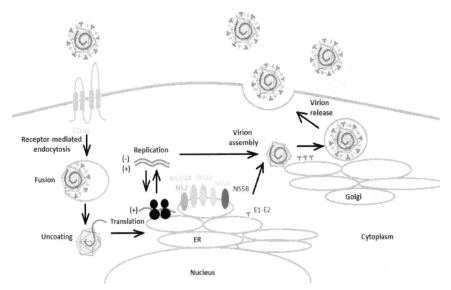

Figure 2. HCV life cycle

2. Models of HCV study

The current standard of HCV therapy that relay on combination with pegylated IFN-α 2a and ribavirin yield limited success rates and reach about 40-80%, depending on HCV genotype, viremia, age and gender of the patients [31-33]. These facts and the large number of infected individuals stress the pressing need for the development of improved antiviral strategies. Understanding of the viral life cycle is crucial to identify effective antiviral agents. The lack of small animal models for HCV hamper studies on viral replication.

2.1. Animal models

The chimpanzee is the only animal that can be infected with HCV. Unfortunately, this model is limited by restricted availability of animals, ethical dilemmas and high cost. Recently, other models have been used to study different aspects of HCV biology and novel antiviral drugs. Researchers had created severe combined immunodeficient (SCID) mice with chimeric livers composed of human and murine hepatocytes that support robust HCV replication [34]. This model was used to demonstrate the antiviral activity of the HCV NS3/4A protease inhibitor BILN-2061. Reduction of the viral load in genotype 1b infected mice after protease inhibitor treatment was similar to results obtained in human clinical trials [35]. Unfortunately, these chimeric mice are not suitable for evaluation of immunotherapeutic agents and vaccines, because of the lack of functional immune system [34].

2.2. Tissue culture

The primary host cell supporting HCV replication is the hepatocyte. HCV replication has been detected in hepatoma, B- and T-cell lines, primary cultures of human or chimpanzee hepatocytes and peripheral blood mononuclear cells. However, the replication levels are very low and do not allow to study HCV replication in detail [36].

Since HCV discovery in 1989, intensive research to create tissue-based model of HCV replication begun. In the late 1990s few independent groups of researchers had developed the first *in vitro* hepatitis C virus replication system based on viral cDNA [37,38]. The HCV genome (called "H77") isolated from patients infected with genotype 1a was used to create plasmid containing full HCV genome transcripted to cDNA. Plasmids containing the full-length HCV cDNA were linearized with *XbaI* and then used as a template in *in vitro* transcription. Transcripted RNAs were infectious, when injected directly into chimpanzees liver [38]. These plasmids were adapted at H77 5′ terminus with the T7 promoter and at 3′ terminus with the hepatitis delta cis-acting ribozyme in continuity with T7 terminator sequences. The vectors were used in transfection of HepG2 and CV-1 cell line. The HCV genomic RNA was generated by T7 RNA polymerase that was provided by infection with recombinant vaccinia virus (vTF7-3) (Figure 3.) [39]. This system has experimental limitations, because of cytopathic and pleiotropic effects of vaccinia virus infection [40].

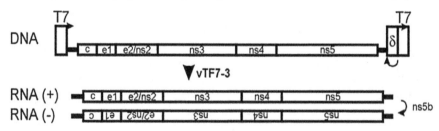

Figure 3. Construction of vectors used for the binary HCV replication system (according to [39], modified)

2.2.1. Development of HCV replicons

According to the low level of HCV replication obtained in full-length cDNA systems, researchers tried to create subgenomic replicon systems. In 1999 the first subgenomic replicon system, which allowed HCV replication in the human hepatoma cell line Huh-7 was established. Viral RNA was isolated from the liver of the patient chronically infected with genotype 1b (Con1). These bicistronic replicons consist of 5′ HCV IRES, neomycin phosphatransferase gene (as a selectable marker), encephalomyocarditis virus (EMCV) IRES and the HCV nonstructural genes from NS2 or NS3 up to NS5B. Translation of the marker was under control of HCV IRES and translation of the HCV polyprotein was under control of EMCV IRES. As a negative control, a defective genome carrying an in-frame 10-amino acid deletion in the NS5B active site was generated (Figure 4).

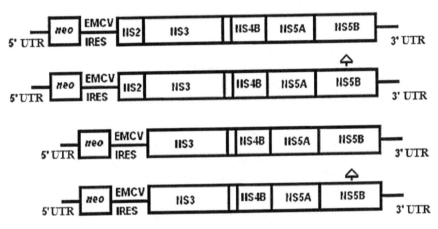

Figure 4. Structure of the HCV subgenomic replicons. The structure of the replicons composed of 5′ HCV UTR (Untranslated Region), neo (neomycin phosphatransferase gene), encephalomyocarditis virus (EMCV) IRES, the HCV nonstructural genes from NS2 or NS3 up to NS5B and 3′ HCV UTR. Δ indicates the position of 10-amino acid deletion in NS5B.

Although transfection of Huh-7 cells with transcripts synthetized *in vitro* and selection with neomycin resulted in a low number of surviving cell colonies, neomycin-resistant cell colonies harbored 1000-5000 copies of positive-sense HCV RNA per cell, which gave evidence of high level replication in transfected cells. To exclude the possibility that resistance was due to plasmid DNA integrated into the host cell genome, DNA of each clone was analyzed by *neo*-specific PCR [41]. Low frequency of transfected cells may indicate that replicon RNAs acquire adaptive mutations to effectively replicate in the Huh-7 cell line or only a low number of cells in the culture support efficient HCV replication. Analysis of the replicon RNAs confirmed the occurrence of cell culture-adaptive mutations that enhance RNA replication [42,43]. Mutations increasing replication were found in the non-structural coding region, especially in NS4B, NS5A and NS5B. The NS3 mutations had minimal or no impact on replication, but can enhance replication synergistically when combined with adaptive mutations in NS4B, NS5A and NS5B. In the same Huh-7 cell line up to 100-fold differences in their ability to support replicon amplification were found, which may indicate that some cellular factors also might be responsible for the different levels of permissiveness of Huh-7 cells [43].

Although HCV is divided into 6 genotypes, replicons have only been reported for genotypes 1 and 2. The establishment of efficiently replicating replicons based on genotype 1a was more challenging than generating functional genotype 1b subgenomic replicons. The first subgenomic replicons (pH77) containing sequences from genotype 1a were constructed in 2003. The replicons were created analogically to previously characterized 1b replicons, and consisted of 5' HCV IRES, *neo*, EMCV IRES and the HCV nonstructural genes from NS2 or NS3 up to NS5B. As a positive control, constructs described in [41] were used. Unfortunately, subgenomic replicons pH77 were unable to support stable replication after RNA electroporation into Huh-7 cells with neomycin selection. Replacing first 75 residues of NS3 coding sequence from type 1a with type 1b resulted in replication. The chimeric subgenomic replicons between HCV type 1a and type 1b were able to replicate in Huh-7 cells, albeit with reduced colony formation efficiency and low viral RNA levels [44]. Transfection of highly permissive Huh-7 subline, Huh-7.5 with H77 replicons containing adaptive Ser-to-Ile substitution (S2204I) in NS5A allowed the development of the first colonies supporting H77 replication. The low frequency of cells supporting H77 replication suggested that efficient H77 replication in Huh-7.5 cells may require at least two adaptive mutations. In all cell clones analyzed, replicating RNAs had acquired a second amino acid substitution in the helicase domain of NS3. Both these mutations, when combined with NS5A S2204I, enhanced the colony-forming ability of subgenomic H77 RNA and allowed the detection of HCV RNA after RNA transfection of either subgenomic replicons [45].

The only non-genotype 1 subgenomic replicon capable of efficient replication in cell culture is the genotype 2a clone JFH1 isolated from a patient with fulminant hepatitis. Fulminant viral hepatitis is a serious form of acute hepatitis and is characterized by a broad viral replication in the host and an intensified host immune response against the virus-infected cells. The replicon was constructed according to the method used in [41], then *in vitro* transcripted and transfected into Huh-7 line. The colony-forming ability of the replicon was 60-fold higher than a Con1 (genotype 1b) subgenomic RNA harboring highly adaptive mutations. Clones of this subgenomic replicon replicated without common amino acid mutations. Furthermore, JFH1 subgenomic replicon replicated efficiently without neomycin selection in a transient replication assay. This genotype 2a subgenomic replicon is important for studying the differences in viral characteristics between genotypes 1 and 2 and to understand the mechanisms of viral replication and persistence [46]. In addition, the JFH1 subgenomic replicon produced colonies in a human hepatocyte-derived cell line, and in IMY-N9, a cell line developed by fusing human hepatocytes and HepG2 cells [47]. Replication of JFH1 subgenomic replicon was also shown in two human non-hepatocyte-derived cell lines, HeLa and 293, which provided useful information about HCV replication and cell tropisms [48].

The subgenomic replicon system is a powerful tool that could be used to study HCV RNA replication in tissue-based assays. It also allows to investigate functions of particular HCV proteins, but has its limitations because of no possibility to product viral particles.

2.2.2. Development of an infectious HCV cell culture system

The first robust cell culture model of HCV infection, in which infectious HCV can be produced, was reported in 2005. To create HCV constructs, HCV RNA was isolated from Japanese patient with fulminant hepatitis C (JFH1) [49]. A 32-year-old man was admitted with a 5-day history of general fatigue, fever and acute liver failure. No evidence of previous liver diseases was found. Anti-HCV antibodies were not detected at the time of admission. The titer of serum HCV RNA was 10^5 copies/ml and the genotype of the isolate was 2a. Its sequence slightly deviates from other genotype 2a strains isolated from patients with chronic hepatitis. This strain has 9,678 bp genome and contains a long open reading frame spanning nucleotide 341-9439 and coding 3033 amino acids [50]. Based on the consensus sequence of JFH1, plasmid pJHF1 containing the full-length JFH1 cDNA downstream of the T7 RNA promoter was constructed (Figure 5).

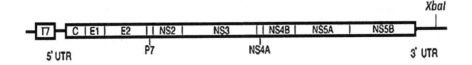

Figure 5. Organization of the full-length HCV construct pJFH1

The plasmids were linearized at the 3′ end of the HCV cDNA by *XbaI* digestion, then *in vitro* transcripted and electroporated to Huh-7. The genome-length RNA was detected in JFH1-transfected cells after 24 h and remained detectable up to 72 h. About 70-80% of the cells were positive for core and nonstructural proteins at 72 h after electroporation, indicating high levels of HCV genome replication in Huh-7 cells. In JFH1-transfected cells the viral RNA concentration in the medium increased rapidly at 5 days after transfection and remained high for the next 7 days, followed by a slow decrease. HCV particles of spherical morphology with an average diameter of about 55 nm were detected in cell medium, which confirms secretion of viral particles after JFH1 transfection into Huh-7. Secreted virus was infectious for naïve Huh-7 cells and for chimpanzee. Although JFH1 virus was infectious for the chimpanzee, infection resulted in transient viraemia and no pathological changes in the liver. Infectivity of viral particles was neutralized by CD81-specific antibodies, confirming specificity of the infection and the important role of CD81 in HCV entry [49].

The JFH1 system described above was limited by low level of infection [49]. Effective viral infection was achieved using cured cell lines such as Huh-7.5 and Huh-7.5.1 [51,52]. These cell lines were obtained by intrerferon treatment of Huh-7 supporting subgenomic replicons. Replicon-containing Huh-7 cells were cured of HCV RNA by initially passing cells twice in the absence of neomycin. Then cells were passaged four times in the presence of IFN-α creating Huh-7.5 line [51]. The Huh-7.5.1 cell line was delivered from Huh-7.5 replicon cell line by culturing 3 weeks in the presence of human IFN-γ to eradicate replicon [52]. Transfection of the JFH1 genome into Huh-7.5 and Huh-7.5.1 cells support high levels of HCV replication in more than 75% of transfected cells. Huh-7.5 line is more permissive for HCV replication than parental Huh-7 [51].

This *in vitro* HCV replication system based on genotype 2a (JFH1) is used worldwide to test the viral life cycle and new antiviral drugs. Its main advantage is markedly enhanced replication efficiency compared with other HCV clones and secretion of virus particles into the cell medium. JFH1 is the only clone without adaptive mutations that is infectious to cultured cells and chimpanzees. The limitation of this method is its restricted adaptation to genotype 2a. The most prevalent and dangerous is infection with genotype 1, which leads to liver damage and hepatocellular carcinoma. It has been difficult to disseminate non-JFH1 HCV strains in cell lines, despite establishment of chimeric viruses. Full-length chimeric genomes were constructed with the use of the core-NS2 gene regions from H77 (genotype 1a) and NS3-NS5B gene regions from JFH1 (genotype 2a). Genotype 1a/2a was able to replicate in Huh-7, but viral particles were not infectious for naïve cells [53]. The JFH1 strain is still the only HCV isolate that can be propagated in Huh-7 cells.

In vitro HCV replication system based on genotype 1a was established in 2006. *In vitro* transcribed full-length HCV RNA from clone H77 was used for transfection of immortalized human hepatocytes (IHH) by electroporation [54]. Human hepatocytes used in this experiment were immortalized by transfection of the HCV core genomic region from genotype 1a [55]. Reverse transcription-PCR of cellular RNA isolated from full-length HCV transfected cells suggested that viral RNA replication appeared. Absence of integrated H77 DNA in IHH genome confirmed HCV genomic RNA replication in the cytoplasm of IHH. The presence of HCV in IHH cell culture medium was detected. Furthermore, virus-like particles were observed in the cytoplasm. HCV infection was also observed after transferring culture media of HCV-replicating cells into naïve IHH [54]. Probably, in IHH cellular defense mechanisms against HCV infection are reduced. Further studies are necessary to test usage of this replication system in novel drug testing.

2.2.3. Cell lines permissive for HCV replication

Hepatoma cell line Huh-7 and its subline Huh-7.5 are the most permissive cell lines for *in vitro* HCV replication identified so far, indicating that a favorable cellular environment exists within these cells. Although adaptive mutations in the HCV NS proteins are required to develop HCV replication at higher frequency. The replication efficiencies of subgenomic RNAs in replication assays can vary by as much as 100-fold between different passages of Huh-7 cells. These differences suggest that effective replication depends on host cell conditions or cellular factors. Replication efficiency decreases with increasing amounts of transfected replicon RNA, indicating that viral RNA or proteins are cytopathic or that host cell factors in Huh7 cells limit RNA replication [43]. Several Huh-7 lines harboring subgenomic HCV replicons were cured of HCV RNA by prolonged treatment with IFN-α. Huh-7.5 is the most permissive cured subline identified so far. The frequency of Huh-7.5 cells able to support HCV replication is approximately three-fold higher than that of the parental Huh-7 cells. Furthermore, more than 75% of the cells that survive the transfection procedure harbor replicating HCV RNAs. The highly permissive subline (Huh-7.5) was obtained from neomycin-selected clones that harbored replicons without adaptive changes in the NS3-5B region [51].

The first *in vitro* non-Huh-7 system of HCV replication was described in 2003. Subgenomic HCV RNAs (Con1) replicated in nonhepatic human epithelial cells. Subgenomic RNA isolated from Huh-7 cell lines that replicate HCV RNA were used in transfection of HeLa cells (cervix carcinoma). Neomycin-resistant cell clones were obtained. Replicons isolated from these cells carried new mutations that could be involved in the control of tropism of the virus [56].

3. Antiviral therapies for chronic hepatitis C

The current standard of treatment of chronic hepatitis C is a combination of pegylated interferon and ribavirin. This therapy leads to 40-50% sustained virological response (SVR) in patients infected with genotype 1, 93% in patients infected with genotype 2, 79% in patients infected with genotype 3 and 69% in patients infected with genotype 4 [29-31]. Among patients infected with genotype 1, only 19-24% treated with PegIFN-α and RBV can achieve rapid virological response (RVR) [57]. Pegylated interferon is associated with numerous side effects: 50% of patients experience flu-like symptoms, 25% psychiatric symptoms, 20% symptoms of fatigue and 10% symptoms of gastritis. The major side effect (36%) of ribavirin is anaemia [58].

Current standard of chronic hepatitis C therapy has limited efficacy and undesirable side effects. There is a pressing need to develop a new antiviral drugs against chronic hepatitis C.

3.1. Inhibitors of hepatitis C virus

An infectious HCV cell culture system is a powerful tool that could be used to study HCV life cycle and function of particular viral proteins. These studies allow to find viral targets for direct-acting antiviral (DDA) drugs and develop Specifically Targeted Antiviral Therapies for HCV (STAT-C). These therapies let to achieve better effectiveness of treatment, its shortening, and the diminishment and limitation of side effects. Current studies are focused on searching the new therapeutic agents for hepatitis C, which are directed against viral proteins. Several HCV inhibitors have reached clinical development, but the most advanced include inhibitors of NS3/4A protease, two of them: boceprevir and telaprevir were approved in 2011 by the Food and Drug Administration (FDA) for the treatment of chronic hepatitis C genotype 1 infection, in combination with PegIFN-α and RBV.

3.1.1. NS3/4A protease inhibitors

The NS3 protease is one of the most attractive targets for developing new therapies against HCV, because of its essential role in viral replication [59,60]. The NS3 inhibitors are divided into two groups, according to different mechanisms of action: covalent and non-covalent inhibitors. Both covalent and non-covalent NS3 protease inhibitors have been developed to bind NS3 active site.

A covalent trap called also 'warhead' form covalent reversible or irreversible bonds with serine hydroxyl of NS3 protease catalytic site [61]. Several classes of HCV protease inhibitors with electrophilic functionality have been reported: aldehyde, ketone, α-ketoamide, α-ketoacid and boric acid/ester. The ketoamides have been the most successful class of covalent inhibitors. Among them, the most clinically advanced are boceprevir and telaprevir [62]. The new AASLD guidelines suggest addition of the NS3/NS4A inhibitors boceprevir or telaprevir to optimize treatment for patients infected with genotype 1.

Boceprevir (SCH 503034) (Figure 6) has been reported as safe and well tolerated in phase I and phase II clinical trials. In a phase II trial (SPRINT-1), patients treated with boceprevir for 48 weeks in combination with pegylated interferon and ribavirin demonstrated SVR rates 67-75% compared with SVR rates of 38% in the control arm [62]. Boceprevir (Victrelis) was approved by the FDA in May 2011, on the basis of the efficacy and safety results from two large phase III clinical studies that evaluated approximately 1,500 adult patients with chronic HCV genotype 1 infection. Both studies included two treatment arms with boceprevir: a response-guided therapy (RGT) arm, in which patients with undetectable virus (HCV-RNA) at week 8 of treatment were eligible for a shorter duration of therapy, as well as a 48-week treatment arm. All patients receiving boceprevir in these studies were first treated with peginterferon alfa-2b and ribavirin in a 4-week lead-in phase, followed by the addition of boceprevir after week 4. The studies also included a control arm in which patients received 48 weeks of treatment with peginterferon alfa-2b and ribavirin alone. In these studies boceprevir yielded sustained virological response rates as high as 67%.

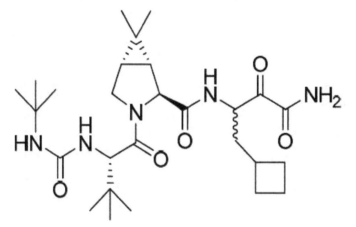

Figure 6. Chemical structure of boceprevir

Boceprevir was approved by FDA for the treatment of chronic hepatitis C genotype 1 infection, in combination with peginterferon alfa and ribavirin, in adult patients with compensated liver disease, including cirrhosis, who were previously untreated or who have failed previous interferon and ribavirin therapy.

Telaprevir (VX-950) (Figure 7) has been reported as generally safe and well tolerated. Its potent inhibition of NS3 protease activity was demonstrated in a mouse model and then proved in a phase II clinical trial [63-65]. In a phase II clinical trial (PROVE-1) observed SVR rate was 35-67% with telaprevir compared with 41% in the control arm. In combination therapy with pegylated interferon and ribavirin (PROVE-2) resulted in 69% SVR [62].

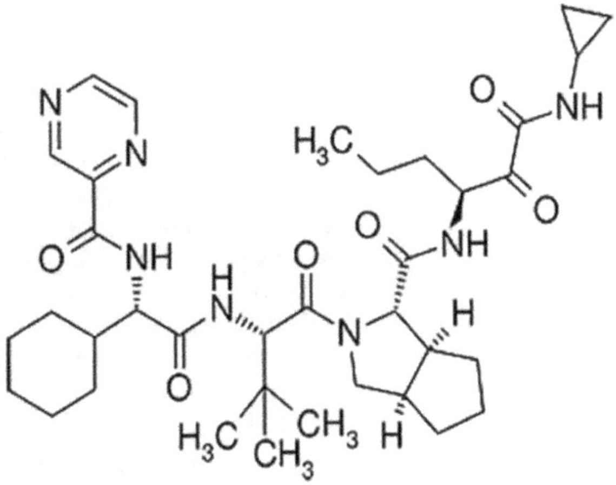

Figure 7. Chemical structure of telaprevir

FDA approved telaprevir (Incivek) in May 2011 to treat certain adults with chronic hepatitis C infection. Telaprevir is used for patients who had either not received interferon-based drug therapy for their infection or who had not responded adequately to prior therapies. Telaprevir is approved for use with combination therapy made up of peginterferon alfa and ribavirin. The safety and effectiveness of telaprevir was evaluated in III phase three clinical trials with about 2,250 adult patients who were previously untreated, or who had received prior therapy. In all studies patients also received the drug with standard of care. In previously untreated patients, 79% of those receiving telaprevir experienced a sustained virologic response (i.e. the infection was no longer detected in the blood 24 weeks after stopping treatment) compared to standard treatment alone. The sustained virologic response for patients treated with telaprevir across all studies, and across all patient groups, was between 20% and 45% higher than current standard of care. The studies indicate that treatment with telaprevir can be shortened from 48 weeks to 24 weeks in most patients. Sixty percent of previously untreated patients achieved an early response and received only 24 weeks of treatment (compared to the standard of care of 48 weeks). The SVR for these patients was 90%.

Several non-covalent inhibitors of NS3 protease have demonstrated potent reduction of HCV RNA and promising safety profiles in HCV infected patients. Non-covalent NS3 protease inhibitors rely on network of hydrogen bonding interactions and polar interactions to obtain binding energy. Ciluprevir was the first direct acting antiviral compound that demonstrated significant viral reduction in patients. The most advanced non-covalent NS3 protease inhibitors: danoprevir, vaniprevir, TMC435350, BI 201335 and BMS-650032 are currently in clinical development [66].

Ciluprevir (BILN-2061) was the first NS3 protease inhibitor tested for antiviral effect in humans. Unfortunately clinical study was terminated due to significant mitochondrial toxicity in cardiac myocytes of several preclinical species [67]. The antiviral efficacy, pharmacokinetics, and tolerability of 25, 200, and 500 mg BILN 2061 twice daily given as monotherapy for 2 days in 31 patients infected with chronic genotype 1 HCV infection and with minimal liver fibrosis were assessed in a placebo-controlled, double-blind pilot study. In 2 subsequent placebo-controlled studies of similar design, 200 mg BILN 2061 twice daily was administered for 2 days to 10 patients with advanced liver fibrosis and to 10 patients with compensated cirrhosis. Viral RNA reductions of 2–3 log10 copies/mL were achieved in most of the patients. There was a trend toward a higher number of patients receiving 500 mg BILN 2061 achieving a viral RNA reduction ≥3 log10 copies/mL as compared with patients receiving 25 mg BILN 2061 [68]. The antiviral activity of ciluprevir was also examined in patients infected with genotype 2 and 3. The antiviral efficacy of BILN-2061 was less pronounced and more variable in patients with HCV genotype 2 or 3 infection compared with previous results in patients with HCV genotype 1 [69].

Antiviral activity of danoprevir (ITMN-191/RG7227) was demonstrated in a randomized, placebo-controlled, 14-day multiple ascending dose study in patients with chronic HCV genotype 1 infection. Danoprevir displayed a slightly more than proportional increase in exposure with increasing daily dose and was rapidly eliminated from the plasma compartment. Maximal decreases in HCV RNA were: -3.9log10IU/ml and -3.2log10IU/ml [70]. Danoprevir is currently in phase IIb clinical study in combination with PegIFN/RBV. Eighty six percent of patients were HCV RNA negative by week 4 of treatment and 92% were negative by week 12 of treatment [66].

Vaniprevir (MK-7009) is another NS3 protease inhibitor, which entered phase IIb development in combination with PegIFN-α and RBV. In the early phase of clinical trial vaniprevir showed 69%-82% clearance of HCV RNA in patients after 4 weeks of treatment.

Opera-1 trial (double blind, placebo-controlled phase IIa trial) examined TMC435350 in combination with PegIFN-α and RBV in patients infected with genotype 1. TMC435350 in combination with PegIFN/RBV showed antiviral activity superior to PegIFN/RBV alone. In the 25, 75, 200 mg 4-week triple therapy arms, 6/9, 9/9 and 10/10 patients had HCV-RNA concentrations below the lower limit of detection (<25 IU/mL) and 3/9, 8/9 and 7/10 had undetectable HCV RNA (<10 IU/mL) at day 28, respectively [71]. TMC435350 is currently in phase IIb clinical trial.

Activity of BI 201335 was demonstrated in phase IIb clinical trial (SILEN-C2) that included HCV infected patients with confirmed non-response to at least 12 weeks of PegIFN/RBV treatment. After 4 weeks of treatment with BI 201335 in combination with PegIFN-α and RBV up to 69% patients had HCV RNA below the limit of detection and after 12 weeks of treatment up to 59% patients had HCV RNA below the limit of detection. SILEN-C2 confirmed robust antiviral activity with overall good tolerability and safety [72].

Asunaprevir (BMS-650032), a novel HCV NS3 protease inhibitor in clinical development, was evaluated for safety, antiviral activity, and resistance in four double-blind, placebo-controlled, sequential-panel, single- and multiple-ascending-dose (SAD and MAD) studies in healthy subjects or patients with chronic HCV genotype 1 infection. Asunaprevir at doses of 200 to 600 mg resulted in rapid HCV RNA decrease from the baseline; maximal mean changes in HCV RNA over time were 2.7 and 3.5 log(10) IU/ml in the SAD and MAD studies, respectively [73]. Currently, asunaprevir is being examined in an innovative programme in combination with NS5A inhibitor BMS-790052 in prior non-responders to PegIFN/RBV therapy [66].

3.1.2. NS3 helicase inhibitors

NS3 helicase is needed for HCV replication and is a potent STAT-C target. Small peptide and tropolones have been reported to inhibit NS3. Peptide inhibitors are quite attractive candidates for antiviral agents. It is relatively easy to design a peptide that fits a studied protein, regardless of the size and chemical properties of the target site. The first experiments performed with a radioactive helicase assay revealed the inhibitory activity of these peptides (of various lengths and composition) and pointed at a peptide composed of 14 amino acids (p14, RRGRTGRGRRGIYR) as the best helicase inhibitor. The first helicase inhibitor corresponded to a highly conserved arginine-rich sequence of domain 2 of the helicase. The 50% inhibitory activity (IC50) value was 725 ± 109 nM, indicating that the peptide is a very efficient NS3 helicase inhibitor. The antiviral activity of p14 was tested in a subgenomic HCV replicon assay that showed that the peptide at micromolar concentrations can reduce HCV RNA replication [74].

Tropolones possess multiple biological activities: antiviral, antimicrobial, and cytotoxic effects on various human tumour cell lines [75-77] . They may also exert an insecticidal as well as a metalloprotease inhibitory effects [78]. The antiviral activity of hydroxylated tropolone derivatives was demonstrated for human influenza virus and human immunodeficiency virus-type 1. Dibromo-morpholinometyltropolone (DBMTr, Figure 8A) is one of the potent NS3 helisace inhibitor among tropolones, which exert anti-helicase activity with an IC50 of 17.56 µM [79].

Other tropolones, like 3,5,7-tri[(4'-methylpiperazin-1'-yl)methyl]tropolone, 3,5,7-tri[(4'-methylpiperidin-1'-yl)methyl]tropolone and 3,5,7-tri[(3'-methylpiperidin-1'-yl)methyl]tropolone demonstrate NS3 inhibition (Figure 8B). Among them, the most active anti-helicase compound 3,5,7-tri[(4'-methylpiperazin-1'-yl)methyl]tropolone (IC50 = 3.4 µM), inhibited RNA replication by 50% at 46.9 µM (EC50) and exhibited the lowest cytotoxicity (CC50) >1 mM resulting in a selectivity index (SI = CC50/EC50) >21. The most efficient replication inhibitor, 3,5,7-tri[(4'-methylpiperidin-1'-yl)methyl]tropolone, inhibited RNA replication with an EC50 of 32.0 µM and a SI value of 17.4, whereas 3,5,7-tri[(3'-methylpiperidin-1'-yl)methyl]tropolone exhibited a slightly lower activity with an EC50 of 35.6 µM and a SI of 9.8. Moreover, these three tropolone derivatives inhibit replication of the HCV subgenomic replicon in cell cultures [80]. Despite the intensive studies, no inhibitors of HS3 helicase are in clinical use.

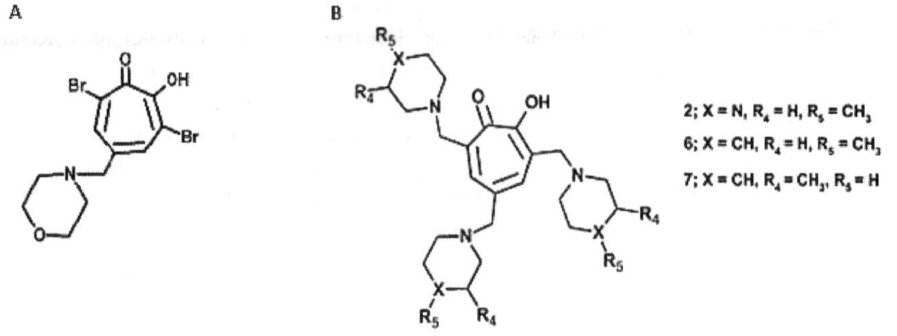

A

B

2; X = N, R₄ = H, R₅ = CH₃

6; X = CH, R₄ = H, R₅ = CH₃

7; X = CH, R₄ = CH₃, R₅ = H

Figure 8. Chemical structure of tropolones: (A): dibromo-morpholinometyltropolone, (B): 3,5,7-tri[(4'-methylpiperazin-1'-yl)methyl]tropolone (2), 3,5,7-tri[(4'-methylpiperidin-1'-yl)methyl]tropolone (6) and 3,5,7-tri[(3'-methylpiperidin-1'-yl)methyl]tropolone (7)

3.1.3. NS5A inhibitors

The HCV NS5A as a part of replication complex is essential for RNA replication. It also has been associated with subverting host intracellular signaling pathways. The potent antiviral activity of different NS5A inhibitors has already been demonstrated in early clinical trial phase. Among them, BMS-790052 entered phase III of clinical trial. Extended Rapid Virologic Response (eRVR) was achieved up to 83% of HCV genotype 1 patients under BMS-790052 treatment in combination with peg-IFN and ribavirin. BMS-790052 was well tolerated with a safety profile comparable in the placebo group [81]. BMS-790052 has been shown *in vitro* to have a low genetic barrier of resistance [82]. Another potent NS5A inhibitor, PPI-461, with a preclinical profile similar to BMS-790052 entered clinical trial [83]. Given the successful nature of clinical trial NS5A inhibitors can now be viewed as a potential cornerstone of HCV combination therapy.

3.1.4. NS5B inhibitors

The HCV NS5B RNA-dependent RNA polymerase as a catalytic component of HCV replication complex plays a crucial role in viral replication cycle. Because NS5B is structurally distinct from mammalian DNA and RNA polymerase enzymes, it is a potent target for the development of new anti-HCV agents. The inhibitors of NS5B are divided into two classes: nucleoside inhibitors (NIs) and non-nucleoside inhibitors (NNIs). The nucleoside inhibitors bind to the active site of polymerase, resulting in chain termination and premature termination of HCV RNA synthesis. Optimization of nucleoside analogues is complicated by NIs interaction with cellular polymerases and signaling pathways. The most advanced candidate for drug is RG7128, which is currently in phase II clinical trial and has shown highly promising safety, tolerability and efficacy profile. In a 14-day monotherapy RG7128 showed high antiviral efficacy, achieving viral load decreases up to 2.7 log IU/ml at the end of dosing period. In phase IIa study 88% of infected patients with genotype 1 and

90% of patients infected with genotype 2 or 3 achieved undetectable viral load after 4 weeks of combination treatment with RG7128, PEG and RBV. Moreover, there were no evidence of resistance selection during two weeks of monotherapy. These results suggest a high antiviral activity in HCV infected patients and a high barrier to resistance of nucleoside analogues as inhibitors of HCV replication [84].

The HCV non-nucleoside inhibitors bind to one of the four allosteric binding sites within the HCV polymerase: site I (Thumb I) for JTK-109, site II (Thumb II) for PF-868554, VCH-759, VCH-916 and VCH-222, site III (Palm I) for ANA-598, A-848837 and ABT-333, and site IV (Palm II) for HCV-796, resulting in conformational changes of the protein and inhibition of catalytic activity of polymerase. Currently, a number of NNIs have entered the clinical trial (Table 1).

Inhibitors	Binding site	Study phase
PF-868554	Site II	Phase II
VCH-759	Site II	Phase I
VCH-916	Site II	Phase I
VCH-222	Site II	Phase I-II
ANA-598	Site III	Phase II
ABT-333	Site III	Phase II
HCV-796	Site IV	Withdrawn

Table 1. HCV NS5B non-nucleoside inhibitors under clinical investigation

HCV-796 was the first inhibitor that showed an antiviral effect in HCV-infected patients, but due to hepatic toxicity HCV-796 was withdrawn from clinical trial. PF-868554, VCH-759, VCH-916, VCH-222, ANA-598 and ABT-333 proved antiviral activity in early clinical trials and some have been advanced to phase II.

PF-868554 was tested in monotherapy over 8 days in genotype 1 patients with chronic infection. All patients demonstrated antiviral response after 48 hours of treatment. In phase IIa clinical trials patients with genotype 1 were treated with PF-868554 in combination with Peg-IFN/RBV. By week 4 HCV RNA plasma levels dropped up to 4.43 $\log_{10}$ compared with placebo and Peg-IFN/RBV group. After 12 weeks of treatment up to 88% had undetectable HCV RNA [85].

VCH-222 antiviral activity in HCV replicons resulted in EC_{50} values of 22.3nM for genotype 1a, 11.2nM for genotype 1b and 4.6μM for genotype 2a. Patients with genotype 1 were treated with VCH-222 for 3 days in phase Ib clinical study. All patients demonstrated rapid and significant antiviral response with 3 $\log_{10}$ in plasma HCV RNA [85].

ANA598 activity against genotype 1 was shown in *in vitro* HCV replicon system [86]. In phase II clinical study genotype 1 patients received ANA598 in combination with Peg-IFN/RBV for 12 weeks. Early virological response was achieved in 73% of HCV infected patients [85].

ABT-333 entered clinical trial after showing in replicon assays its potent inhibition of genotype 1a and 1b polymerases. Antiviral activity of ABT-333 was evaluated in combination with Peg-IFN/RBV. At the last day of therapy (Day 28) 41.7% of patients had undetectable HCV RNA [85].

Non-nucleoside polymerase inhibitors selected different NS5B mutations which exhibited resistance profiles. Given the distinct binding sites and resistance profiles among different NNIs, it is likely that two to three NNIs could be used in combination [85]. Antiviral activity and tolerability of the HCV NNIs is promising, but the use of NS5B inhibitors in treatment of HCV infection need more studies in future clinical trials.

4. Conclusions

The current standard of care (SOC) of chronic hepatitis C infection based on pegylated-IFN and ribavirin has limited efficacy and undesirable side effects. Efficient therapies must be developed to eliminate HCV infections, which pose a serious worldwide health problem. More than 80% of HCV infected patients develop the persistent infection, which leads to serious liver diseases, like liver cirrhosis or hepatocellular carcinoma. The lack of small animal models for HCV hamper studies on viral replication and search for potent antiviral targets. Development of HCV replicons able to replicate in cell line Huh-7 and *in vitro* HCV infection system using the JFH-1 clone provide a good method for screening the new antiviral drugs. An infectious HCV cell culture system is a powerful tool that could be used to study HCV life cycle and function of particular viral proteins. These studies allow to find viral targets for direct-acting antiviral (DAA) drugs and develop Specifically Targeted Antiviral Therapies for HCV (STAT-C). STAT-C let to achieve better effectiveness of treatment, its shortening, and the diminishment and limitation of side effects. Current studies are focused on searching for the new therapeutic agents for hepatitis C, which are directed against viral enzymes. Several HCV pro-drugs selected in *in vitro* HCV replicon system as   potent HCV inhibitors, have reached clinical development, but the most advanced include inhibitors of NS3/4A protease. Two of them: boceprevir and telaprevir were approved in 2011 by the Food and Drug Administration (FDA) for the treatment of chronic hepatitis C genotype 1 infection, in combination with PegIFN-α and RBV.

The most intensive studies are focused on searching for novel DAAs against HCV genotype 1. Infections with genotype 1 are the most prevalent and dangerous, leading to liver injury and hepatocellular carcinoma. Moreover, patients infected with genotypes 1 respond to SOC less effectively than patients infected with genotypes 2 and 3. Because an infectious HCV cell culture system is restrictive adapt to genotype 2a, subgenomic HCV replicon systems are used worldwide for screening new antivirals against genotype 1.

In conclusion, *in vitro* tissue-based model of HCV replication allowed to determine the viral life cycle and function of particular viral proteins, which is crucial for the development of novel more efficient antivirals.

Author details

Paulina Godzik

National Institute of Public Health – National Institute of Hygiene,
Department of Virology, Warsaw, Poland,

Acknowledgement

This work was supported by grant NN 405 132 539 "Tissue culture studies on potent drugs against HCV" from Polish Ministry of Science and Higher Education. The author is grateful to prof. Kazimierz Madaliński for reviewing the manuscript.

5. References

[1] WHO (2000) Hepatitis C-global prevalence (update). Wkly. Epidemiol. Rec. 75:18-9.

[2] Saito I, Miyamura T, Ohbayashi A, Harada H, Katayama T, Kikuchi S, Watanabe Y, Koi S, Onji M, Ohta Y, Choo QL, Houghton M, Kuo G (1990) Hepatitis C virus infection is associated with the development of hepatocellular carcinoma. Proc. Natl. Acad. Sci. USA 87: 6547-6549.

[3] Lohmann V, Körner F, Koch J, Herian U, Theilmann L, Bartenschlager R (1999) Replication of subgenomic hepatitis C virus RNAs in a hepatoma cell line. Science 285:110-113.

[4] Choo QL, Richman KH, Han JH, Berger K, Lee C, Dong C, Gallegos C, Coit D, Medina-Selby R, Barr PJ, Weiner A J, Bradley DW, Kuo G, Houghton M (1991) Genetic organization and diversity of the hepatitis C virus. Proc. Natl. Acad. Sci. USA. 88: 2451-2455.

[5] Bukh J, Purcell RH, Miller RH (1992) Sequence analysis of the 5' noncoding region of hepatitis C virus. Proc. Natl. Acad. Sci. USA. 89: 4942-4946.

[6] Honda M, Beard MR, Ping LH, Lemon SM (1999) A phylogenetically conserved stem-loop structure at the 5' border of the internal ribosome entry site of hepatitis C virus is required for cap-independent viral translation. J. Virol. 73: 1165-1174.

[7] Tanaka T, Kato N, Cho MJ, Shimotohno K (1995) A novel sequence found at the 3' terminus of hepatitis C virus genome. Biochem. Biophys. Res. Commun. 215: 744-749.

[8] Kolykhalov AA, Feinstone SM, Rice CM (1996) Identification of a highly conserved sequence element at the 3' terminus of hepatitis C virus genome RNA. J. Virol. 70: 3363-3371.

[9] Martell M, Esteban JI, Quer J, Genescà J, Weiner A, Esteban R, Guardia J, Gómez J (1992) Hepatitis C virus (HCV) circulates as a population of different but closely related genomes: quasispecies nature of HCV genome distribution. J. Virol. 66: 3225-3229.

[10] Dusheiko G, Schmilovitz-Weiss H, Brown D, McOmish F, Yap PL, Sherlock S, McIntyre N, Simmonds P (1994) Hepatitis C virus genotypes: an investigation of type-specific differences in geographic origin and disease. Hepatology. 19: 13-18.

[11] Bruno S, Silini E, Crosignani A, Borzio F, Leandro G, Bono F, Asti M, Rossi S, Larghi A, Cerino A, Podda M, Mondelli MU (1997) Hepatitis C virus genotypes and risk of hepatocellular carcinoma in cirrhosis: a prospective study. Hepatology. 25: 754-758.

[12] Zein NN (2000) Clinical significance of hepatitis C virus genotypes. Clin. Microbiol. Rev.13: 223-35.

[13] Zein NN, Rakela J, Krawitt EL, Reddy KR, Tominaga T, Persing DH (1996) Hepatitis C virus genotypes in the United States: epidemiology, pathogenicity, and response to interferon therapy. Collaborative Study Group. Ann. Intern. Med. 125: 634-639.

[14] Ray RB, Lagging LM, Meyer K, Steele R, Ray R (1995) Transcriptional regulation of cellular and viral promoters by the hepatitis C virus core protein. Virus. Res. 37: 209-220.

[15] Chou AH, Tsai HF, Wu YY, Hu CY, Hwang LH, Hsu PI, Hsu PN (2005) Hepatitis C virus core protein modulates TRAIL-mediated apoptosis by enhancing Bid cleavage and activation of mitochondria apoptosis signaling pathway. J. Immunol. 174: 2160-2166.

[16] Nielsen SU, Bassendine MF, Burt AD, Bevitt DJ, Toms GL (2004) Characterization of the genome and structural proteins of hepatitis C virus resolved from infected human liver. J. Gen. Virol. 85: 1497-507.

[17] Pileri P, Uematsu Y, Campagnoli S, Galli G, Falugi F, Petracca R, Weiner AJ, Houghton M, Rosa D, Grandi G, Abrignani S (1998) Binding of hepatitis C virus to CD81. Science. 282: 938-941.

[18] Flint M, McKeating JA (2000) The role of the hepatitis C virus glycoproteins in infection. Rev. Med. Virol. 10: 101-117.

[19] Griffin SD, Beales LP, Clarke DS, Worsfold O, Evans SD, Jaeger J, Harris MP, Rowlands DJ (2003) The p7 protein of hepatitis C virus forms an ion channel that is blocked by the antiviral drug, Amantadine. FEBS Lett. 535: 34-38.

[20] Pavlović D, Neville DC, Argaud O, Blumberg B, Dwek RA, Fischer WB, Zitzmann N (2003) The hepatitis C virus p7 protein forms an ion channel that is inhibited by long-alkyl-chain iminosugar derivatives. Proc. Natl. Acad. Sci. USA. 100: 6104-6108.

[21] Grakoui A, McCourt DW, Wychowski C, Feinstone SM, Rice CM (1993) A second hepatitis C virus-encoded proteinase. Proc. Natl. Acad. Sci. USA. 90:10583-10587.

[22] Suzuki T, Ishii K, Aizaki H, Wakita T (2007) Hepatitis C viral life cycle. Adv. Drug. Deliv. Rev. 59:1200-1212.

[23] Gretton SN, Taylor AI, McLauchlan J (2005) Mobility of the hepatitis C virus NS4B protein on the endoplasmic reticulum membrane and membrane-associated foci. J. Gen. Virol. 86:1415-1421.

[24] Wünschmann S, Medh JD, Klinzmann D, Schmidt WN, Stapleton JT (2000) Characterization of hepatitis C virus (HCV) and HCV E2 interactions with CD81 and the low-density lipoprotein receptor. J. Virol. 74: 10055-10062.

[25] Bartosch B, Vitelli A, Granier C, Goujon C, Dubuisson J, Pascale S, Scarselli E, Cortese R, Nicosia A, Cosset FL (2003) Cell entry of hepatitis C virus requires a set of co-receptors

that include the CD81 tetraspanin and the SR-B1 scavenger receptor. J. Biol. Chem. 278: 41624-41630

[26] Saunier B, Triyatni M, Ulianich L, Maruvada P, Yen P, Kohn LD (2003) Role of the asialoglycoprotein receptor in binding and entry of hepatitis C virus structural proteins in cultured human hepatocytes. J. Virol. 77: 546-559.

[27] Voisset C, Callens N, Blanchard E, Op De Beeck A, Dubuisson J, Vu-Dac N (2005) High density lipoproteins facilitate hepatitis C virus entry through the scavenger receptor class B type I. J. Biol. Chem. 280: 7793-7799.

[28] Chevaliez S, Pawlotsky JM (2006) HCV genome and life cycle. In: Tan SL, editor. Hepatitis C viruses: Genomes and Molecular Biology. Norfolk (UK): Horizon Bioscience. pp.5-47.

[29] Tanaka Y, Shimoike T, Ishii K, Suzuki R, Suzuki T, Ushijima H, Matsuura Y, Miyamura T (2000) Selective binding of hepatitis C virus core protein to synthetic oligonucleotides corresponding to the 5' untranslated region of the viral genome. Virology. 270: 229-236.

[30] Klein KC, Dellos SR, Lingappa JR (2005) Identification of residues in the hepatitis C virus core protein that are critical for capsid assembly in a cell-free system. J. Virol. 79: 6814-6826.

[31] Zeuzem S, Hultcrantz R, Bourliere M, Goeser T, Marcellin P, Sanchez-Tapias J, Sarrazin C, Harvey J, Brass C, Albrecht J (2004) Peginterferon alfa-2b plus ribavirin for treatment of chronic hepatitis C in previously untreated patients infected with HCV genotypes 2 or 3. J Hepatol. 40: 993-999.

[32] Kamal SM, El Tawil AA, Nakano T, He Q, Rasenack J, Hakam SA, Saleh WA, Ismail A, Aziz AA, Madwar MA (2005) Peginterferon {alpha}-2b and ribavirin therapy in chronic hepatitis C genotype 4: impact of treatment duration and viral kinetics on sustained virological response. Gut. 54: 858-866.

[33] Hoofnagle JH, Seeff LB. Peginterferon and ribavirin for chronic hepatitis C. N. Engl. J. Med. 355: 2444-2451.

[34] Mercer DF, Schiller DE, Elliott JF, Douglas DN, Hao C, Rinfret A, Addison WR, Fischer KP, Churchill TA, Lakey JR, Tyrrell DL, Kneteman NM (2001). Hepatitis C virus replication in mice with chimeric human livers. Nat. Med. 7: 927-933.

[35] Meuleman P, Libbrecht L, De Vos R, de Hemptinne B, Gevaert K, Vandekerckhove J, Roskams T, Leroux-Roels G (2005) Morphological and biochemical characterization of a human liver in a uPA-SCID mouse chimera. Hepatology. 41: 847-856.

[36] Blight KJ, Norgard EA (2006) HCV replicon systems. In: Tan SL, editor. Hepatitis C viruses: Genomes and Molecular Biology. Norfolk (UK): Horizon Bioscience. pp. 311-351.

[37] Kolykhalov A.A., Agapov E.V., Blight K.J., Mihalik K., Feinstone S.M., Rice C.M (1997) Transmission of hepatitis C by intrahepatic inoculation with transcribed RNA. Science. 277: 570-574.

[38] Yanagi M., Purcell R.H., Emerson S.U., Bukh J (1997) Transcripts from a single full-length cDNA clone of hepatitis C virus are infectious when directly transfected into the liver of a chimpanzee. Proc. Natl. Acad. Sci. USA. 9: 8738-8743.

[39] Chung RT, He W, Saquib A, Contreras AM, Xavier RJ, Chawla A, Wang TC, Schmidt EV (2001) Hepatitis C virus replication is directly inhibited by IFN-alpha in a full-length binary expression system. Proc. Natl. Acad. Sci. USA. 98: 9847-9852.

[40] Pietschmann T, Bartenschalager R (2003) Tissue culture and animal models for hepatitis C virus. Clin. Liver. Dis. 7: 23-43.

[41] Lohmann V, Korner F, Koch JO, Herian U, Theilmann L, Bartenschlager R (1999) Replication of subgenomic hepatitis C virus RNAs in a hepatoma cell line. Science. 285: 110-113.

[42] Blight KJ, Kolykhalov AA, Rice CM (2000) Efficient initiation of HCV RNA replication in cell culture. Science. 290: 1972-1974.

[43] Lohmann V, Hoffmann S, Herian U, Penin F, Bartenschlager R (2003) Viral and cellular determinants of hepatitis C virus RNA replication in cell culture. J. Virol. 77: 3007-3019.

[44] Gu B, Gates AT, Isken O, Behrens SE, Sarisky RT (2003) Replication studies using genotype 1a subgenomic hepatitis C virus replicons. J. Virol. 77: 5352-5359.

[45] Blight KJ, McKeating JA, Marcotrigiano J, Rice CM (2003) Efficient replication of hepatitis C virus genotype 1a RNAs in cell culture. J. Virol. 77: 3181-3190.

[46] Kato T, Date T, Miyamoto M, Furusaka A, Tokushige K, Mizokami M, Wakita T (2003) Efficient replication of the genotype 2a hepatitis C virus subgenomic replicon. Gastroenterology. 125: 1808-1817.

[47] Date T, Kato T, Miyamoto M, Zhao Z, Yasui K, Mizokami M, Wakita T (2004) Genotype 2a hepatitis C virus subgenomic replicon can replicate in HepG2 and IMY-N9 cells. J. Biol. Chem. 279: 22371-22376.

[48] Kato T, Date T, Miyamoto M, Zhao Z, Mizokami M, Wakita T (2005) Nonhepatic cell lines HeLa and 293 support efficient replication of the hepatitis C virus genotype 2a subgenomic replicon. J. Virol. 79: 592-596.

[49] Wakita T, Pietschmann T, Kato T, Date T, Miyamoto M, Zhao Z, Murthy K, Habermann A, Kräusslich HG, Mizokami M, Bartenschlager R, Liang TJ (2005) Production of infectious hepatitis C virus in tissue culture from a cloned viral genome. Nat. Med. 11: 791-796.

[50] Kato T, Furusaka A, Miyamoto M, Date T, Yasui K, Hiramoto J, Nagayama K, Tanaka T, Wakita T (2001) Sequence analysis of hepatitis C virus isolated from a fulminant hepatitis patient. J. Med. Virol. 64: 334-339.

[51] Blight KJ, McKeating JA, Rice CM (2002) Highly permissive cell lines for subgenomic and genomic hepatitis C virus RNA replication. J. Virol. 76: 13001-13014.

[52] Zhong J, Gastaminza P, Cheng G, Kapadia S, Kato T, Burton DR, Wieland SF, Uprichard SL, Wakita T, Chisari FV (2005) Robust hepatitis C virus infection in vitro. Proc. Natl. Acad. Sci. USA. 102: 9294-9299.

[53] Lindenbach BD, Evans MJ, Syder AJ, Wölk B, Tellinghuisen TL, Liu CC, Maruyama T, Hynes RO, Burton DR, McKeating JA, Rice CM (2005) Complete replication of hepatitis C virus in cell culture. Science. 309: 623-626.

[54] Kanda T, Basu A, Steele R, Wakita T, Ryerse JS, Ray R, Ray RB (2006) Generation of infectious hepatitis C virus in immortalized human hepatocytes. J. Virol. 80: 4633-4639.

[55] Basu A, Meyer K, Ray RB, Ray R (2002) Hepatitis C virus core protein is necessary for the maintenance of immortalized human hepatocytes. Virology. 298: 53-62.

[56] Zhu Q, Guo JT, Seeger C (2003) Replication of hepatitis C virus subgenomes in nonhepatic epithelial and mouse hepatoma cells. J. Virol. 77: 9204-9210.

[57] Kanda T, Imazeki F, Yokosuka O (2010) New antiviral therapies for chronic hepatitis C. Hepatol. Int. 4: 548-561.

[58] Lee LY, Tong CY, Wong T, Wilkinson M (2012) New therapies for chronic hepatitis C infection: a systematic review of evidence from clinical trials. Int. J. Clin. Pract. 66: 342-55.

[59] Lamarre D, Anderson PC, Bailey M, Beaulieu P, Bolger G, Bonneau P, Bös M, Cameron DR, Cartier M, Cordingley MG, Faucher AM, Goudreau N, Kawai SH, Kukolj G, Lagacé L, LaPlante SR, Narjes H, Poupart MA, Rancourt J, Sentjens RE, St George R, Simoneau B, Steinmann G, Thibeault D, Tsantrizos YS, Weldon SM, Yong CL, Llinàs-Brunet M (2003) An NS3 protease inhibitor with antiviral effects in humans infected with hepatitis C virus. Nature. 426: 186-189

[60] Perni RB, Almquist SJ, Byrn RA, Chandorkar G, Chaturvedi PR, Courtney LF, Decker CJ, Dinehart K, Gates CA, Harbeson SL, Heiser A, Kalkeri G, Kolaczkowski E, Lin K, Luong YP, Rao BG, Taylor WP, Thomson JA, Tung RD, Wei Y, Kwong AD, Lin C (2006) Preclinical profile of VX-950, a potent, selective, and orally bioavailable inhibitor of hepatitis C virus NS3-4A serine protease. Antimicrob. Agents Chemother. 50: 899-909.

[61] Lin C, Kwong AD, Perni RB (2006) Discovery and development of VX-950, a novel, covalent, and reversible inhibitor of hepatitis C virus NS3.4A serine protease. Infect. Disord. Drug Targets. 6: 3-16.

[62] Chen KX, Njoroge FG (2011) NS3 protease covalent inhibitors. In:Tan S-L, He Y, editors. Hepatitis C. Antiviral drug discovery and development. Norfolk (UK): Caister Academic Press. pp.169-192.

[63] McHutchison JG, Everson GT, Gordon SC, Jacobson IM, Sulkowski M, Kauffman R, McNair L, Alam J, Muir AJ; PROVE1 Study Team (2009) Telaprevir with peginterferon and ribavirin for chronic HCV genotype 1 infection. N. Engl. J. Med. 360: 1827-1838.

[64] Hézode Ch, Foretier N, Dusheiko G, Ferenci P, Pol S, Goeser T, Bronowicki JP, Bourlière M, Gharakhanian S, Bengtsson L, McNair L, George S, Kieffer T, Kwong A, Kauffman RS, Alam J, Pawlotsky JM, Zeuzem S; PROVE2 Study Team (2009) Telaprevir and peginterferon with or without ribavirin for chronic HCV infection. N. Engl. J. Med. 360: 1839-1850.

[65] McHutchison JG, Manns MP, Muir AJ, Terrault NA, Jacobson IM, Afdhal NH, Heathcote EJ, Zeuzem S, Reesink HW, Garg J, Bsharat M, George S, Kauffman RS, Adda

N, Di Bisceglie AM; PROVE3 Study Team (2010) Telaprevir for previously treated chronic HCV infection. N. Engl. J. Med. 362: 1292-303.

[66] Buckman BO, Kossen K, Nicholas JB, Seiwert SD (2011) NS3 protease non-covalent inhibitors. In: Tan S-L, He Y, editors. Hepatitis C. Antiviral drug discovery and development. Norfolk (UK): Caister Academic Press. pp. 194-214.

[67] Vanwolleghem T, Meuleman P, Libbrecht L, Roskams T, De Vos R, Leroux-Roels G (2007) Ultra-rapid cardiotoxicity of the hepatitis C virus protease inhibitor BILN 2061 in the urokinase-type plasminogen activator mouse. Gastroenterology. 133: 1144-1155.

[68] Hinrichsen H, Benhamou Y, Wedemeyer H, Reiser M, Sentjens RE, Calleja JL, Forns X, Erhardt A, Crönlein J, Chaves RL, Yong CL, Nehmiz G, Steinmann GG (2004) Short-term antiviral efficacy of BILN 2061, a hepatitis C virus serine protease inhibitor, in hepatitis C genotype 1 patients. Gastroenterology. 127: 1347-1355.

[69] Reiser M, Hinrichsen H, Benhamou Y, Reesink HW, Wedemeyer H, Avendano C, Riba N, Yong CL, Nehmiz G, Steinmann GG (2005) Antiviral efficacy of NS3-serine protease inhibitor BILN-2061 in patients with chronic genotype 2 and 3 hepatitis C. Hepatology. 41: 832-835.

[70] Forestier N, Larrey D, Guyader D, Marcellin P, Rouzier R, Patat A, Smith P, Bradford W, Porter S, Blatt L, Seiwert SD, Zeuzem S (2011) Treatment of chronic hepatitis C patients with the NS3/4A protease inhibitor danoprevir (ITMN-191/RG7227) leads to robust reductions in viral RNA: a phase 1b multiple ascending dose study. J. Hepatol. 54: 1130-1136.

[71] Manns M, Reesink H, Moreno C, Berg T, Benhamou Y, Horsmans Y, Dusheiko G, Flisiak R, Meyvisch P, Lenz O, Sekar V, van't Klooster G, Simmen K, Verloes R (2009) Opera-1 trial: interim analysis of safety and antiviral activity of TMC435 in treatment-naive genotype 1 HCV patients. J. Hepatol. 50 (Suppl. 1): S7

[72] Sulkowski M, Bourliere M, Bronowicki JP, Streinu-Cercel A, Preotescu L, Asselah T, Pawlotsky JM, Shafran S, Pol S, Caruntu FA, Mauss S, Larrey D, Häfner C, Datsenko Y, Stern J, Kubiak R, Steinmann G. (2010) SILEN-C2: early antiviral activity and safety of BI 201335 combined with peginterferon alfa-2a and ribavirin (PEGIFN/RBV) in chronic HCV genotype-1 patients with non-response to PEGIFN/RBV. J. Hepatol. 52 (Suppl. 1): S462-S463

[73] Pasquinelli C, McPhee F, Eley T, Villegas C, Sandy K, Sheridan P, Persson A, Huang SP, Hernandez D, Sheaffer AK, Scola P, Marbury T, Lawitz E, Goldwater R, Rodriguez-Torres M, Demicco M, Wright D, Charlton M, Kraft WK, Lopez-Talavera JC, Grasela DM (2012) Single- and multiple-ascending-dose studies of the NS3 protease inhibitor asunaprevir in subjects with or without chronic hepatitis C. Antimicrob. Agents Chemother. 56: 1838-1844.

[74] Gozdek A, Zhukov I, Polkowska A, Poznanski J, Stankiewicz-Drogon A, Pawlowicz JM, Zagórski-Ostoja W, Borowski P, Boguszewska-Chachulska AM (2008) NS3 Peptide, a

novel potent hepatitis C virus NS3 helicase inhibitor: its mechanism of action and antiviral activity in the replicon system. Antimicrob. Agents Chemother. 52: 393-401.

[75] Miyamoto D, Kusagaya Y, Endo N, Sometani A, Takeo S, Suzuki T, Arima Y, Nakajima K, Suzuki Y (1998) Thujaplicin-copper chelates inhibit replication of human influenza viruses. Antiviral Res. 39: 89–100.

[76] Inamori Y, Shinohara S, Tsujibo H, Okabe T, Morita Y, Sakagami Y, Kumeda Y, Ishida N (1999) Antimicrobial activity and metalloprotease inhibition of hinokitiol-related compounds, the constituents of Thujopsis dolabrata S. and Z. hondai MAK. Biol. Pharm. Bull. 22: 990–993.

[77] Budihas SR, Gorshkova I, Gaidamakov S, Wamiru A, Bona MK, Parniak MA, Crouch RJ, McMahon JB, Beutler JA, Le Grice SF (2005) Selective inhibition of HIV-1 reverse transcriptase-associated ribonuclease H activity by hydroxylated tropolones. Nucleic Acids Res.33: 1249–1256.

[78] Piettre SR, Andre C, Chanal MC, Ducep JB, Lesur B, Piriou F, Raboisson P, Rondeau JM, Schelcher C, Zimmermann P, Ganzhorn AJ (1997) Monoaryl- and bisaryldihydroxytropolones as potent inhibitors of inositol monophosphatase. J. Med. Chem. 40: 4208–4221.

[79] Boguszewska-Chachulska AM, Krawczyk M, Najda A, Kopańska K, Stankiewicz-Drogoń A, Zagórski-Ostoja W, Bretner M (2006) Searching for a new anti-HCV therapy: synthesis and properties of tropolone derivatives. Biochem. Biophys. Res. Commun. 341: 641-647.

[80] Najda-Bernatowicz A, Krawczyk M, Stankiewicz-Drogoń A, Bretner M, Boguszewska-Chachulska AM (2010) Studies on the anti-hepatitis C virus activity of newly synthesized tropolone derivatives: identification of NS3 helicase inhibitors that specifically inhibit subgenomic HCV replication. Bioorg. Med. Chem. 18: 5129-5136.

[81] Pol S, Everson G, Ghalib R, Rustgi V, Martorell C, Tatum HA, Lim J, Hezode C, Diva U, Yin PD, Hindes R (2010) Once-daily NS5A inhibitor (BMS-790052) plus peginterferon-alpha-2A and ribavirin produces high rates of extended rapid virologic response in treatment-naive HCV-genotype 1 subjects: phase 2a trial. J. Hepatol. 52 (Suppl. 1): S462.

[82] Gao M, Nettles RE, Belema M, Snyder LB, Nguyen VN, Fridell RA, Serrano-Wu MH, Langley DR, Sun JH, O'Boyle DR 2nd, Lemm JA, Wang C, Knipe JO, Chien C, Colonno RJ, Grasela DM, Meanwell NA, Hamann LG (2010) Chemical genetics strategy identifies an HCV NS5A inhibitor with a potent clinical effect. Nature. 465: 96-100.

[83] Colonno R, Peng E, Bencsik M, Huang N, Zhong M, Huq A, Huang Q, Williams J, Li L (2010) Identification and characterization of PPI-461, a potent and selective HCV NS5A inhibitor with activity against all HCV genotypes. J. Hepatol. 52: S14-15.

[84] Klumpp K, Smith M (2011) Nucleoside inhibitors of hepatitis C virus. In: Tan S-L, He Y, editors. Hepatitis C. Antiviral drug discovery and development. Norfolk (UK): Caister Academic Press. pp. 293-309.

[85] Thompson P, Patel R, Steffy K, Appleman J (2009) Preclinical studies of ANA598 combined with other anti-HCV agents demonstrate potential of combination treatment. J. Hepatol. 50 (Suppl. 1): S37.

Chapter 12

Viral DNA and cDNA Array in the Diagnosis of Respiratory Tract Infections

B. Matteoli and L. Ceccherini-Nelli

Additional information is available at the end of the chapter

1. Introduction

1.1. Respiratory Tract Infections (RTIs)

Respiratory tract infections (RTIs) are caused directly or indirectly by any infectious agent that implants and replicates in the respiratory tract, in the pulmonary parenchyma or onto pleural sierose and causes clinical syndromes with prevalent respiratory symptoms [1]. When the respiratory tract is only the first site of infection, patients show respiratory syndromes, but then symptoms involve the specific anatomic district target of the infectious agent, as we can see in some infectious disease: Measles, Scarlet fever, Mononucleosis, Meningococcal disease and Varicella. Even with these limitations RTIs are the most common infections in men and are a significant cause of morbidity and mortality in both developing and developed countries among infants, youngsters and elderly people, and are the first cause of temporary invalidation (absence from work or school), visiting emergency service and family doctor consultancies during the winter season, independently from the age [2]. so they are a great socioeconomic and medical burden. In 2010, the World Health Organization (WHO) estimated that RTIs caused about 3 million deaths worldwide, including developed countries, being the first cause of child mortality [3]. Respiratory infections are common in both hospital and community settings. The third national prevalence survey conducted in 2006 found that infections of the lower respiratory tract (LRTI, not pneumonia) and pneumonia together accounted for 19.9 percent of the Healthcare Associated Infections (HCAIs) in acute hospitals. It I important also to consider the hospital-acquired infections affecting the respiratory tract that cause considerable morbidity and mortality. This type of respiratory infections generally affects those who are affected from serious diseases [4]. RTIs can be classified in, infections of the upper respiratory tract (URIs), which affect the nose, sinuses and throat (common cold, tonsillitis, sinusitis, laryngitis, influenza) and infections of the lower respiratory tract (LRTIs), which affect the airways and lungs (influenza, bronchitis, pneumonia, bronchiolitis, tuberculosis, that is a persistent bacterial infection of the lungs). LRTIs include two serious conditions – acute bronchitis and pneumonia:

Acute bronchitis (inflammation of the bronchi) is an acute respiratory infection in which the dominant symptom is coughing without localized infection. It must not not be confused with chronic bronchitis, which is a chronic obstructive pulmonary disease (COPD). Acute bronchitis is usually an infection that is community-acquired and typically it arises as a complication of URI caused by a virus, when bacterial infection supervenes. Children that seem prone to bronchitis generally have poor living conditions (overcrowding,poor hygiene and poor nutrition) and the respiratory disease may be exacerbated by maternal smoking, especially during pregnancy. It I reported that individuals who have experienced childhood bronchitis are at risk of developing further symptoms during their teenage years if they then smoke [5]. Pneumonia (inflammation of the lung) is a serious condition, that caused may death after RTI, especially in older adults and infants. It may be acquired in hospital or the community. The alveoli become filled with pus, air is excluded, and the lung is said to be 'consolidated'.In bronchopneumonia, consolidation is widely distributed; in lobar pneumonia,it is localized [6]. In the community, bacterial pneumonia is most frequently caused by *Streptococcus pneumonia* that infects most commonly people with pre-existinghealth problems, frequently developing as a complication of some other RTI (for example influenza or measles). The establishment of a pharmacological treatment is complicated because some strains of *Streptococcus pneumoniae* are now resistant to penicillin, so vaccination has been recommended in the UK since 2003. The pneumococcal vaccine is part of the childhood immunization program and it is also recommended to people over 65 years of age. It is also recommended for people following splenectomy and those with dysfunction of the spleen, sickle cell disease, coeliac disease, chronic renal disease, chronic respiratory disease, chronic heart conditions, liver disease, diabetes mellitus, immunosuppression and HIV. After vaccination, about 80 per cent of healthy adults develop a good antibody response within three weeks. Other bacteria responsible for community-acquired pneumonia include *Mycoplasma pneumoniae, Haemophilus influenzae, Legionella pneumophila* and *Staphylococcus aureus*, including the strain that produces the Panton–Valentine leukocidin toxin [7]. It I mandatory also to consider hospital acquired pneumonia. It mainly affects critically ill and postoperative patients. Risk factors include obesity, impaired consciousness, a history of smoking and underlying respiratory disease. In hospital, bacteria, viruses or fungi can cause pneumonia, but most hospital-acquired pneumonia is caused by *S. aureus* and Gram-negative opportunists [8]. URIs that involve the nasal passages, pharynx, tonsils and epiglottis are minor infections acquired in the community and are caused by viruses. URIs can, however, have serious consequences for the very young and older adults. They also account for a high proportion of days lost from work and school, so their impact on the health of individuals and their social and economic consequences should not be dismissed. The nasal discharge associated with colds contains viral particles, dead cells from the nasal mucosa and bacteria, but these are of the same type as are present in health. Bacterial invasion of the damaged epithelium is rare, and antibiotics are seldom required. Acute ear infection occurs as a complication with up to 30 percent of URIs. Because most URIs are self-limiting, their complications are more important than the infections, mastoiditis and other complications of URIs account for nearly 5 percent of all URI worldwide leading to hearing impairment or deafness most of the times in developing countries where there is limited access to adequate medical treatments. RTIs are extraordinary frequent because of a great number of antigenically distinct aetiological agents, their great diffusion and the short period of immunization. Moreover the anatomic structure of the respiratory tract shows high variable physicochemical parameters such as temperature, pH, and humidity that assurance a great number of habitat for various microorganisms. Moreover the respiratory tract is directly connected to the external environment and it is continuously crossed by the airflow that frequently contains irritant agents (atmospheric contaminants, cigarette smoke, cold air, fines) that may injure the local mucosa end predispose to the implantation of microorganisms. Fortunately there is a huge amount of mechanisms of defense that preserve the integrity of this anatomic site, the nasopharyngeal lymphatic system, ciliary cells, salivary lisozima, antibodies, interferons and pulmonary macrophafges. The infectious agents of RTIs can be viruses, bacteria, fungi and protozoa, but they are usually caused by a virus [1, 2]. Bacterial agents causing RTIs can implant directly onto the mucosa of the respiratory tract, but more frequently they superinfect a tract of mucosa previously injured by a virus. *Streptococcus pneumoniae*, *Haemophilus influenzae*, and *Moraxella catarrhalis* are the most common organisms that cause the bacterial superinfection of viral acute sinusitis [8]. The sequence viral-bacterial infection is very frequent and must be considered to establish an appropriate pharmacological treatment. Parasitic infections of the respiratory tract occur worldwide among both immunocompetent and immunocompromised patients and may affect the respiratory system in a variety of ways. The most common parasites involved are, *Ascariasis*, *Schistosoma* and *Toxoplasma gondii*. Since the clinical presentations and radiographic findings of several of these diseases may mimic tuberculosis and malignancy it is important to consider parasitic infections in the differential diagnosis of such respiratory syndromes. If identified early, most parasitic respiratory diseases are curable with medical or surgical treatments [9]. RTIs are transmitted by airflow, pathogens enter the organism trough the upper respiratory tract and sometimes trough the conjunctivae, that have been infected by direct contact especially for viruses, or trough vectors, whereas the elimination of the infecting particles takes place trough cough. Rarely the respiratory tract can be reached by circulating microorganisms in blood. The reservoir for the majority of RTIss are infected humans that guarantee an optimal environment for the survival of the infecting agent, but some microorganisms that occasionally infect humans have animal reservoir; for example in 2002, a new Coronavirus (CoV) emerged in the People's Republic of China, associated with a severe acute respiratory syndrome (SARS) and mortality in humans. The epidemic rapidly spread throughout the world before being contained in 2003, although sporadic cases occurred thereafter in Asia. The virus was thought to be of zoonotic origin from a wild animal reservoir (Himalayan palm civets, *Paguma larvata*), but the definitive host is still unknown [10]. RTIs are generally epidemic because they are very contagious, generally they are more frequent during autumn and winter when atmospheric pollution is higher and permanence indoor increases. Acquired immunity after RTIs is generally short because the majority of the most common aetiological agents are ineffective immunogens; the immunological response to RTIs generally produces initially IgM and after IgG, so that the presence of IgG has a diagnostic value only if it increases for times or is very high. During RTIs many viral and bacterial immunogens induce the production of secretory IgA that are released in to the muco where they compete for tissue receptors and link to microorganisms. The periodic genetic mutations of RTIs aetiological agents vanish both natural and acquired immunity; each time that genetic mutations that involve major antigens take place, especially for viruses, the diffusion of the infecting agent of RTIs becomes pandemic [11]. An example of immune evasion due to genetic variability is given by INFVs that are dynamic and are continuously evolving. INFVs can change in two different ways: antigenic drift and antigenic shift. Antigenic drift takes place continuously while antigenic shift happens only occasionally. INF A viruses undergo both kinds of changes; INF B viruses change only by the more gradual process of antigenic drift. Antigenic drift refers to small, gradual changes that occur through point mutations in the two genes that contain the genetic material to produce the main surface proteins, hemagglutinin (HA), and neuraminidase (NA). These point mutations occur unpredictably and result in minor changes to these surface proteins. Antigenic drift produces new virus strains that may not be recognized by antibodies to earlier INFVs strains. This is one of the main reasons why people can become infected with INFVs more than one time and why global surveillance is critical in order to monitor the evolution of human INFVs for selection of which strains should be included in the annual production of INF vaccine. In most years, one or two of the three virus strains in the INF vaccine are updated to keep up with the changes in the circulating INFVs. For this reason, the immunization against INF needs to be vaccinated every year. Antigenic shift refers to an abrupt, major change to produce a novel INFV A subtype in humans that was not currently circulating among people. Antigenic shift can occur either through direct animal (poultry)-to-human transmission or through mixing of human influenza A and animal influenza A virus genes to create a new human influenza A subtype virus through genetic reassortment. Antigenic shift results in a new human INFV A subtype [12, 13].

2. Viral RTIs

At least two hundred different viruses can establish RTIs and they belong to the *Adenoviridae, Orthomyxoviridae, Paramyxoviridae, Picornaviridae, Coronaviridae, Herpesviridae* and *Parvoviridae* familes. Viruses the most frequently associated with RTIs are, Adenovirus (ADV), BoV, CoV, Enterovirus (ENTV), INFV A, B, C, Metapneumovirus (hMetV), Parainfluenza (IPV) viruses 1, 2, 3, 4, Rhinoviruses (RV), Respiratory Syncitial Viruses (RSV). While it is true that respiratory viruses place a greater burden on people in developing countries, these viruses are still a big health threat in the developed world, where over 100 million people have been killed INFV in the last century alone (Piralla et al. 2011). In children, viruses are responsible for the majority of RTIs, with bacteria thought to be responsible for fewer than 15% of cases; acute pharyngitis is caused by viruses in more than 70% percent of cases in young children, mild pharyngeal redness and swelling and tonsil enlargement are typical [2, 15]. The most pathogenetic viruses for humans are, INFV and RSV with high mortality rate in elderly people and infants respectively. INFVs have the highest evolution rate, being the INF A viruses those causing the most severe and expansive outbreaks. Genetic variations in INF A viruses usually lead to global epidemics or pandemics; the latest FluA (H1N1) outbreak in 2009 was originated by a variant INF A H1N1 of swine origin, classified by the World Health Organization (WHO) as level 6 alert, pandemic [16]. Viruses cause most URIs, with RVs accounting for the 25-30% of cases, IPV, ADV, RSV, INFV 25-35%, CoV 10% and ENTV (Coxsackievirus, CoxV) less than 5% [17]. Acute viral infections predispose to bacterial infections of the sinuses and middle ear, and aspiration of infected secretions and cells can result in LRIs. To date the most common causes of viral LRIs are RSVs. They tend to be highly seasonal, unlike PINFV, the next most common cause of viral LRIs. The epidemiology of influenza viruses in children in developing countries deserves urgent investigation because safe and effective vaccines are available. Before the effective use of measles vaccine, the Measles virus was the most important viral cause of respiratory tract–related morbidity and mortality in children in developing countries. ADVs are medium-sized (90-100 nm), non-enveloped icosohedral viruses with double-stranded DNA. There are over 50 types that are immunologically distinct that can cause infections in humans. ADVS are relatively resistant to chemical and physical agents and to adverse pH conditions and can live for a long time outside the body and most commonly cause respiratory illness. The symptoms of ADV infection can range from the common cold to pneumonia, croup, and bronchitis. Some ADVs types can cause other illnesses such as gastroenteritis, conjunctivitis, cystitis, and less commonly, neurological disease. Infants and people with weakened immune systems are at high risk for severe complications of ADV infection and some people infected with ADV can have ongoing infections in their tonsils, adenoids, and intestines that do not cause symptoms. They can shed the virus for months or years [18].Human bocavirus (hBoV), the second parvovirus potentially pathogenic to humans after Parvovirus B19 (Pb19), was discovered by PCR in respiratory samples collected from young children with respiratory diseases in Sweden in 2005. Since the first description of hBoV as a possible human pathogen of lower respiratory tract infections in children, it has been detected in at least 19 countries in the five continents. HBoV infections shows a variety of clinical symptoms; the most common symptoms in hBoV-infected children without coinfections are cough (85%), followed by rinorrhea (67%), fever (59%), difficulty in breathing (48%), diarrhea (16%), conjunctivitis (9%) and rash (9%), body temperature ranging from 37.5 to 40.2°C, wheezing; nausea, sore throat, headache and myalgia were also recorded in hBoV-infected children of older age and adults. The age distribution of hBoV-infected humans ranges from 10 days to 60 years, but hBoV was primarily detected in young children aged 6 months to 3 years. That a peak detection of hBoV is among children of 6 to 24 months of age [19]. CoVs (order Nidovirales, family Coronaviridae, genus Coronavirus) are a diverse group of large, enveloped, positive-stranded RNA viruses that cause respiratory and enteric diseases in humans and other animals. There are three groups of CoV; groups 1 and 2 contain mammalian viruses, while group 3 contains only avian viruses. Within each group, CoVs are classified into distinct species by host range, antigenic relationships, and genomic organization. The viruses can cause severe disease in many animals, and several viruses, including infectious bronchitis virus, feline infectious peritonitis virus, and transmissible gastroenteritis virus, are significant veterinary pathogens. Human coronaviruses (hCoVs) are found in both group 1 (HCoV-229E) and group 2 (HCoV-OC43) and are responsiblefor ~30% of mild upper respiratory tract illnesses [19]. In March 2003, a novel CoV, SARS-CoV was discovered in association with cases of severe acute respiratory syndrome (SARS). The sequence of the complete genome of SARS-CoV has been now determined; it is 29,727 nucleotides in length, has 11 open reading frames, and the genome organization is similar to that of other CoVs. Phylogenetic analyses and sequence comparisons showed that SARS-CoV is not closely related to any of the previously characterized CoV. By late April 2003, over 4300 SARS cases and 250 SARS-related deaths were reported to WHO from over 25 countries around the world. Most of these cases occurred after exposure to SARS patients in household or healthcare settings. The incubation period for the disease is usually from 2 to 7 days. Infection is usually characterized by fever, which is followed a few days later by a dry, non-productive cough, and shortness of breath. Death from progressive respiratory failure occurs in about 3% to nearly 10% of cases. Evidence of SARS-CoV infection has been documented in SARS patients throughout the world. SARSCoV RNA has frequently been detected in respiratory specimens, and convalescent-phase serum specimens from SARS patients contain antibodies that react with SARS-CoV [19]. Enteroviruses (ENTV) genus belongs to the *Picornaviridae* family positive single stand ssRNA viruses. Current taxonomy divides non-polio human ENTVs into four species (human ENTVs A to D), including a total of 108 serotypes. Individual serotypes have different temporal patterns of circulation and can be associated with different clinical manifestations. Although the majority of human ENTV infections remain asymptomatic, these viruses are associated with various clinical syndromes, ranging from minor febrile illness to severe and potentially fatal pathologies, including aseptic meningitis, encephalitis, myopericarditis, acute flaccid paralysis, and severe neonatal sepsis-like disease. ENTVs are responsible for a wide range URIs and LRTIs occurring in adults and infants. These viruses are considered as the third etiological cause of bronchiolitis in young infants aged 1-12 months. Moreover, several clinical case studies reported the etiological role of the Cox A16, the ENTV 71 and of a newly discovered genotype ENTV-104 in the development of acute or fatal pneumonia indicating that ENTVs belonging to species A to C can be responsible for severe LRTIs in immunocompetent infants or adults. Taking into account the recent epidemiological and clinical data and because of frequent mutations and intra-species enteroviral RNA genomic recombination events, the respiratory strains of ENTV are considered also as potential agents of emerging infectious diseases in human populations [21]. There are three types of INFV: A, B, and C. Only INF A viruses are further classified by subtype on the basis of the two main surface glycoproteins HA and NA. INF type A viruses can infect people, birds, pigs, horses, and other animals, but wild birds are the natural hosts for these viruses. INF viruses type A are divided into subtypes and named on the basis of two proteins on the surface of the virus: HA and NA. There are 16 known HA subtypes and 9 known NA subtypes and many different combinations of HA and NA proteins are possible. Only some INF A subtypes (i.e., H1N1, H1N2, and H3N2) are currently in general circulation among people, other subtypes are found most commonly in other animal species. For example, H7N7 and H3N8 viruses cause illness in horses, and H3N8 also has recently been shown to cause illness in dogs. Only INF A viruses infect birds, and all known subtypes of INF A viruses can infect birds. Typically, wild birds do not become sick when they are infected with avian INF A viruses, however, domestic poultry, such as turkeys and chickens, can become very sick and die from avian flu, and some avian INF A viruses also can cause serious disease and death in wild birds. Highly pathogenic avian INF A virus strains (HPAI) can cause severe illness and high mortality in poultry. More recently, some HPAI viruses (e.g., H5N1) have been found to cause no illness in some poultry, such as ducks. Avian INF A viruses of the subtypes H5 and H7,including H5N1, H7N7, and H7N3 viruses, have been associated with HPAI, and human infection with these viruses have ranged from mild (H7N3, H7N7) to severe and fatal disease (H7N7, H5N1). In general, direct human infection with avian INFVs occurs very infrequently, and has been associated with direct contact (e.g., touching) infected sick or dead infected birds (domestic poultry). INF B viruses are usually found only in humans. Unlike INF A viruses, these viruses are not classified according to subtype. INF B viruses can cause morbidity and mortality among humans, but in general are associated with less severe epidemics than INF A viruses. Although INF type B viruses can cause human epidemics, they have not caused pandemics. INF C viruses cause mild illness in humans and do not cause epidemics or pandemics. These viruses are not classified according to subtype [22].

HMPV is a respiratory viral pathogen that causes a spectrum of illnesses that range from asymptomatic infection to severe bronchiolitis. In 2001, van den Hoogen et al described the identification of this new human viral pathogen from respiratory samples submitted for viral culture during the winter season. Half of the initial 28 hMPV isolates were cultured from patients younger than 1 year, and 96% were isolated from children younger than 6 years. Seroprevalence studies revealed that 25% of all children aged 6-12 months who were tested in the Netherlands had detectable antibodies to hMPV; by age 5 years, 100% of patients showed evidence of past infection. Separate reports from all areas of the world support the early contention that this newly discovered virus is ubiquitous, and, like human respiratory RSV infection, is seasonal in nature. Although the description of this viral pathogen was first described in children, subsequent reports have highlighted the importance of hMPV as a cause of respiratory illness in adults of all ages in patients with cancer, in the elderly population (as a cause of serious LRTI), and in adults with underlying chronic medical conditions [23]. Human parainfluenza viruses (hPIV) belong to the *Paramyxoviridae* family and are negative-sense, single-stranded RNA viruses that show fusion and hemagglutinin-neuraminidase glycoprotein "spikes" on their surface. There are four serotypes (1 through 4) and two subtypes (4a and 4b) that show different clinical and epidemiologic features. The virion varies in size (average diameter between 150 and 200 nm) and shape, is unstable in the environment (surviving a few hours on environmental surfaces), and is readily inactivated with soap and water. HPIV are common causes of RTIs in infants and young children; the most distinctive clinical feature of HPIV-1 and HPIV-2 is croup (i.e., laryngotracheobronchitis or swelling around the vocal chords and other parts of the upper and middle airway); HPIV-1 is the leading cause of croup in children, whereas HPIV-2 is less frequently detected. HPIV-3 is more often associated with bronchiolitis (swelling of the small airways leading to the lungs) and pneumonia. HPIV-4 is detected infrequently, and is less likely to cause severe disease; but it may be more common than once thought. HPIVs can cause repeated infections with all serotypes throughout life. Reinfections usually manifested by an upper respiratory tract illness (e.g., a cold, sore throat). HPIVs can also cause serious LRTIs with repeat infection (e.g., pneumonia, bronchitis, and bronchiolitis), especially among older adults and patients with compromised immune systems. The incubation period (time from exposure to the virus to onset of symptoms) for HPIVs generally ranges from 2 to 7 days [24]. RVs are small (30 nm), nonenveloped viruses that contain a single-strand RNA genome within an icosahedral (20-sided) capsid, that RV can be transmitted by aerosol or direct contact. The nasal mucosa is the primary site of onculation, although the conjunctiva may also be involved, though to a lesser extent. RVs attache their selves to the respiratory epithelium and spreads locally. The optimum temperature for RVs repolication is 33-35°C and so does not replicate efficiently at body temperature. This could be the major reason why RVs replicate well in the nasal passages and upper trachebronchial tree but not so well in the lower respiratory tract. RVs are well known for causing the common cold, although they have also been implicated in causing bronchitis and asthma attacks. There is little or no cross-protection between serotypes, making it very difficult to make vaccines. These viruses seems to affect children first, and then there are many modes of transmission that range from aerosol to direct hand-to-hand contact [25]. RSV, is a respiratory virus negative-sense, single-stranded RNA of the family *Paramyxoviridae* that infects the lungs and breathing passages. Most otherwise healthy people recover from RSV infection in 1 to 2 peeks. However, infection can be severe in some people, such as certain infants, young children, and older adults. In fact, RSV is the most common cause of bronchiolitis (inflammation of the small airways in the lung) and pneumonia in children under 1 year of age in the United States. In addition, RSV is more often being recognized as an important cause of respiratory illness in older adults. Initially isolated RSV from chimpanzees with URI as the causative agent of most epidemic bronchiolitis cases. Subsequently, RSV has been associated LRTIs infection in infants and multiple epidemiologic studies have confirmed the role of this virus as the leading cause of LRT infection in infants and young children. Peak of incidence of occurrence of severe RSV disease is observed at age 2-8 months. Overall, 4-5 million children younger than 4 years acquire an RSV infection, and more than 125,000 children are hospitalized annually in the United States because of this infection. This translates to 3-9 per 1000 children younger than 1 year who are hospitalized annually for this condition. Virtually all children have had at least one RSV infection by their third birthday. The WHO has targeted RSV for vaccine development, which is not surprising, given the prevalence and potential severity of this condition [26]. Other less common cause of viral URI is Herpes simplex virus (HSV).

Table 1 resumes viral cusative agents of Respiratory Tract Infection (RTIs).

Virus group	Antigenic types	RTIs
Rhinoviruses	100 types and 1 subtypes	Common Cold, Pharyngitis, Acute Laryngitis, Sinusitis, Acute Bronchitis, Bronchiolitis, Acute Pneumonia
Coronavirus	3 or more types	Common Cold, Pharyngitis, Acute Laryngitis, Acute Bronchitis
Parainfluenza viruses	4 types	Common Cold, Pharyngitis, Acute Laryngitis, Acute Laryngotracheobronchitis (Croup), Sinusitis, Acute Bronchitis, Bronchiolitis, Acute Pneumonia
Respiratory Syncytial virus	2 types	Common Cold, Acute Laryngotracheobronchitis (Croup), Acute Bronchitis, Bronchiolitis, Acute Pneumonia
Influenza viruses	3 types	Common Cold, Pharyngitis, Acute Laryngitis, Acute Laryngotracheobronchitis (Croup), Sinusitis, Acute Bronchitis, Bronchiolitis, Acute Pneumonia
Adenoviruses	47 types	Common Cold, Pharyngitis, Acute Laryngitis, Sinusitis, , Acute Bronchitis, Bronchiolitis, Acute Pneumonia
Human Metapneumovirus		Common Cold, , Acute Laryngitis, Bronchiolitis, Acute Pneumonia
Rubeola virus		Common Cold, Acute Bronchitis, Acute Pneumonia
Enteroviruses	5 types	Common cold, Pharyngitis, Acute Bronchitis, Bronchiolitis, Acute Pneumonia
Rubella virus		Common cold, , Acute Bronchitis
Varicella Zooster virus		Common Cold, Oral cavity Infections, Acute Pneumonia
Herpes Simplex viruses	2 types	Pharyngitis, Epiglottitis, Oral cavity Infections, Acute Pneumonia
Cytomegalovirus		Pharyngitis, Oral cavity Infections, Acute Pneumonia
Human Immunodeficency virus	1 type	Pharyngitis
Epstein Barr virus		Epiglottitis
Human herpes virus 6		Acute Pneumonia

Table 1. Viral cusative agents of Respiratory Tract Infection (RTIs)

The relative importance of individual viral agents in early life is open to debate. Certainly, RSV, RV, PIV, and INFV are predominant in the published data. However there are several factors limiting the ability to draw a definitive conclusion about which virus is the most common or important: differences in the way that data are collected (PCR versus immunoassay) between and within studies and the impact of assay sensitivity; differences in study design affecting age, recruitment criteria, and which viruses are studied. About viral aetiology and infant hospitalization due to respiratory infection, INFVs, ADV, hMPV, PIV, RVs, and RSV can all cause bronchiolitis, necessitating hospitalization; RSV has most commonly been reported to be the main cause of hospitalization due to bronchiolitis and increased disease severity, followed by RV and then by influenza virus and viral coinfection is relatively common, occurring in about 20% of cases. Even there is no consensus on the effect of coinfection on disease severity, coinfection with both hMPV and RSV increased the intensive care unit admission rate. While knowledge of which virus is predominant is relevant for the design of vaccines and specific prophylactic treatments, what can be observed is the similarity of symptoms caused by a wide range of viral agents. It may therefore be more appropriate to focus on ways to target the symptoms and not the agent. This may be especially relevant when an excess immune response causes the disease or when there are multiple serologically distinct subtypes circulating.

3. Diagnosis of viral RTIs

Respiratory viruses, that belong to several taxonomic families, show overlapping clinical signs and symptoms. In clinical practice, a specific virus is often not identified due to the lack of sensitive tests and/or the presence of as-yet-unknown pathogens [27]. Because of the great variety of possible pathogenic agents involved, and because of the high frequency of coinfections, especially among young children, whose immune system is still developing, it is mandatory to use diagnostic methods that allow multiple, sensitive, efficacious and rapid identification of all possible viruses simultaneously possibly present in the clinical sample [27]. Respiratory viruses have become increasingly recognized as serious causes of morbidity and mortality in immunocompromised patients. Rapid and sensitive detection of respiratory viruses is essential for early diagnosis and administration of appropriate antiviral therapy as well as for effective implementation of infection control measures.

Rapid diagnosis of respiratory viruses can enable:

- To establish a direct antiviral therapy, when available, that is crucial considering that antivirals are only effective if administered in the early stages of infection ;
- To abolish unnecessary use of antibiotics, that are often prescribed to patients infected with respiratory viruses, with the result in no relief from symptoms and likelihood that antibiotic resistance will occur in concomitant bacteria ;
- To understand the viral natural history and pathophysiology, which may allow physicians to better understand potential complications[28];
- To implement appropriate personal protective equipment and measures, such as quarantine of infected patients, to minimize spread and prevention of unnecessary isolation (often at great expense) of uninfected individuals; particularly important with

newly emerging or re-emerging pathogens, including severe acute respiratory syndrome caused by CoV, highly pathogenic avian INF and swine-origin INFV H1N1 (S-OIV H1N1) [29];

- To perform accurate epidemiological studies, that allow clinicians to identify populations at risk and determine which populations should consider vaccination (if a suitable vaccine is available);

- finally, to implement rapid viral diagnosis that significantly decreases length of hospital stay and unnecessary laboratory testing [30].

However, even with the best viral detection assays currently available, a specific pathogen cannot be identified in 20% to 50% of RTIss. Existing viral diagnostic methods are limited in sensitivity and scope [31]. Several works suggest that respiratory viruses are underdiagnosed and they might be responsible for a considerable part of the total number of non characterized acquired pneumonia cases [32, 33, 34, 35, 36]. Traditional diagnostics methods besides being too slow, laborious and with a low sensitivity threshold, do not identify common viruses and high incidence viruses, like RV, or new viruses, like CoV. Emerging viruses like MetV or BoV recently discovered and with a very high clinical incidence, especially among children, are not detected neither with the use of traditional techniques nor with more modern methods of molecular diagnostics[37]. The discovery of respiratory viruses occurred initially between 1933 and 1965 when INFV, ENTV, ADV, RSV, RV, PINV and CoV were found by virus culture. In the 1990s, the development of high throughput viral detection and diagnostics instruments increased diagnostic sensitivity and enabled the search and the discover also of new viruses [38]. Since many respiratory viruses can present with similar signs and symptoms, it is impossible to differentiate one virus infection from another clinically. The clinician therefore relies on the laboratory to identify the virus. Many clinicians commonly diagnose patients syndromically with influenza or influenza-like illness without laboratory identification of a virus. The positive predictive value (PPV) of a clinical diagnosis of influenza virus infection in an adult case ranged from 18% to 87% compared with cases of laboratory-confirmed influenza virus infection. During periods of high INF virus activity, a clinical diagnosis based on acute onset of high fever and cough can be highly predictive of influenza (PPV, 79% to 87%; negative predictive value [NPV], 39% to 75%). The consequences of not identifying INF virus in a nursing home or on a hospital ward could be catastrophic. INFVs outbreaks in a hospital can be devastating given the wide range of immunocompromised patients (cancer patients and transplant recipients) that are highly susceptible to life-threatening influenza virus infection. In either setting, specific antiviral agents such as M2 channel inhibitors (amantidine and rimantidine) or NA inhibitors (oseltamivir and zanamavir) can be prescribed; however, these drugs are effective only when given within the first 24 h following infection. The traditional methods by which respiratory viruses are routinely diagnosed: virus culture, serology, immunofluorescence/antigen detection, and nucleic acid/PCR-based tests. The gold standard for the diagnosis of respiratory virus infections, virus isolation by culture, takes days to weeks (shell vial assays are not available for the diagnostics of all respiratory viruses), and many new viruses remain unculturable [39]. Virus culture consists in infecting cell lines with clinical samples; virus isolation is performed using three or four cell lines and, together with embryonated hen eggs for INFV, provides the means for isolating respiratory viruses. Tissue cultures can take up to 15 days, therefore, the infection can often be resolved before the infectious agent is defined. Shell vial culture (SVC), first described in the early 1990s for murine Cytomegalovirus (CMV) is a modification of the conventional cell culture technique for rapid detection of viruses *in vitro* that involves inoculation of the clinical specimen on to cell monolayer grown on a cover slip in a shell vial culture tube, followed by low speed centrifugation and incubation. In this system the low speed centrifugation enhances viral infectivity to the susceptible cells because it produces a minor trauma to the cell surface by the low speed centrifugation mechanical force and enhances the viral entry in to the cells, reducing the total time taken for the virus to produce infection of cells. Shell vials of R-Mix, a combination of mink lung cells and human adenocarcinoma cells (strains Mv1Lu and A549) enable the detection of respiratory viruses from prospective clinical respiratory specimens [40]. The rapidity of the technique without any compromise on sensitivity has made SVC very popular in the field of clinical virology [41]. SVC as traditional cell culture assay requires specific technical and manual skills and is performed only in specialized laboratories. Isolation in cell culture, with the traditional method or the SVC assay is time consuming and have low sensitivity, but it is still considered the gold standard for the diagnosis of viral infections because molecular methods may not necessarily indicate that the virus is causing disease, viral RNA has been detected in asymptomatic children and during viral persistence. For these reasons traditional cell culture and SVC assay are still routinely used diagnostically and have a predominant role in epidemiological studies, are used to follow the course of an infection. A variety of serological tests including the hemagglutination inhibition (HAI) test, complement fixation, and enzyme immunoassay (EIA) are used for testing paired acute- and convalescent-phase sera for diagnosing infections, and in the case of INFV, HAI is able to subtype the virus as being H1 or H3 virus. EIAs was introduced in the 1980s and 1990s, lacks in sensitivity and is usually relegated to point-of-care testing in defined settings. CFT and haeagglutination-inhibition HAI techniques are usually used for serology. Any serological diagnosis is going to be retrospective because the antibody response to a viral infection can take 2 weeks to develop, but serological tests as cell culture assays are used in epidemiological studies. Serology assays that test blood samples for either virus-specific antibodies or viral antigen by a functional assay are labor-intensive and slow to produce results. Direct fluorescent antibody (DFA) staining of cells derived from nasopharyngeal swabs or nasopharyngeal aspirates (NPA) became the mainstay for many laboratories and provide a rapid test result in about 3 h. Direct fluorescent antibody (DFA) but suffers from low sensitivity and is available for only a limited number of respiratory viruses [42]. Molecular tests are more sensitive than other diagnostic approaches, including virus isolation in cell culture, SVC), DFA staining, and EIA, and now form the backbone of clinical virology laboratory testing around the world. The polymerase chain reaction (PCR) is a scientific technique in molecular biology to amplify a single or a few copies of a piece of DNA or cDNA across several orders of magnitude, generating thousands to millions of copies of a particular sequences. Reverse transcription-PCR (RT-PCR), that produces a cDNA template from an RNA template and amplify the cDNA target is a highly sensitive method for diagnosis of viral infection and has been used successfully in children with RSV. This method was found to be 100-fold more sensitive than single-round PCR and was capable of detecting 0.05 PFU of tissue culture-passaged virus. Multiplex polymerase chain reaction (PCR) assays introduced in the last ten years to avoid separate amplifications of the viruses under investigation that are resource intensive, time consuming and labor intensive and can detect up to 19 different viruses in a single test using numerous primer couples that have the same annealig temperature. Several multiplex PCR tests are now commercially available and tests are working their way into clinical laboratories, but the majority of the multiplex PCR assays have not included recently discovered respiratory pathogens and require validation of results by post PCR hybridization or semi/nested PCR which make the assay cost ineffective and increases chances of cross contamination. The appearance of eight new respiratory viruses, including the SARS CoV in 2003 and swine-origin INF A/H1N1 in 2009, in the human population in the past nine years has tested the ability of virology laboratories to develop diagnostic tests to identify these viruses. Nucleic acid amplification procedures including PCR, nucleic acid sequence-based amplification (NASBA), and loop-mediated isothermal amplification (LAMP) were developed for most respiratory viruses by the end of the decade, and today, these highly sensitive NAATs are used in the routine clinical laboratory for detecting respiratory viruses. The profile of viruses detected in RTIs is changing due to the increasing use of nucleic acid-based diagnostic screens and the discovery of newly isolated viruses. Knowledge of the infecting agent does not routinely alter treatment except insofar as a positive viral identification will reduce the inappropriate use of antibiotics and may allow the cohorting of patients to reduce nosocomial infection. Several "new" viruses have been characterized, in part triggered by especially RT-PCR. Recently isolated respiratory viral agents include human hMPV, found in samples from children with RSV-like bronchiolitis who were RSV negative; hBoV, discovered by a random PCR screen of respiratory tract samples; and two new polyomaviruses, WU and KI. The discovery of new agents of infection is important because they may play a role as coinfecting agents, altering disease severity. Newly discovered viruses may also be important in future outbreaks; for example, the severe acute respiratory syndrome (SARS) caused by a CoV [43]. Polymerase chain reaction (PCR) testing is rapid and highly sensitive. At times it seems that it has supplanted culture isolation as a new gold standard for the detection of respiratory viruses in a research setting. However, most PCR tests target only 1 virus at a time, making these assays cumbersome when screening a clinical specimen for all viruses that have a PCR test available. Moreover, reliable PCR assay need to be developed to detect or identify novel viruses. Molecular technology has better sensitivity and the development of multiplex amplifications makes it possible to detect a broader panel of viruses. ADV, PIV, hCoV, BoV can now be detected by multiplex assays. These assays are based on different types of technology, such as ligation-dependent probe amplification (MLPA1), dual priming oligonucleotide (DPO) technology, target specific primer extension (TSPE), or target-specific extension (TSE). For the simultaneous detection of up to 20 viruses, a number of multiplex PCR assays have been proposed [44, 45, 46, 47, 48]. It is believed that PCR has replaced tissue culture and serology as the gold standard for the detection of respiratory viruses owing to its speed, availability and versatility. Even if molecular detection has many proven advantages over standard virological methods, tissue culture remains an important method for detecting novel viral mutations within a virus population, for the detection of novel viruses and for phenotypic characterization of viral isolates [48]. Recently, DNA microarray testing has emerged as a promising new technology for broad-spectrum virus detection [49, 50, 51]. Panviral DNA microarrays represent the most robust approach for massively parallel viral surveillance and detection. The Virochip (Virochip; University of California San Francisco

[UCSF]) is a panviral DNA microarray capable of detecting all known viruses, as well as novel viruses related to known viral families in a single assay. The Virochip has been used to indentify SARS, Xenotropic murine leukemia virus-related (a novel Retrovirus) from patients with familial prostate cancer, and a novel clade of human RV [52]. The Virochip has also proven to be successful in a clinical veterinary setting by successfully identifying a novel CoV from a beluga whale held in an aquatic containment facility and by identifying foot-and-mouth disease virus (FMDV) in ticks collected from a livestock market in Nairobi, Kenya. However, the usefulness and sensitivity of the Virochip platform have not been tested on clinical veterinary specimens [53]. In-house microarray platforms have been designed to detect all known viruses, as well as novel viruses related to known viral families (Virochip; University of California San Francisco [UCSF]). These Virochip consists of 22 000 oligonucleotide probes representing all 1800 fully or partially sequenced viruses in GenBank as of Fall 2004 [54]. The performance of the Virochip in respiratory virus detection has been tested using virally infected tissue culture cells and in selected patient cohorts, and it demonstrated high sensitivity and specificity [55]. To date, the Virochip has not been compared directly with standard diagnostic tests for viruses in a clinical setting; thus, the Virochip has been compared with conventional clinical DFA- and PCR-based testing in the detection of respiratory viruses associated with RTIs in children [56]. We report In this study the analysis of 10 clinical samples from patients, with respiratory tract infection symptoms that resulted negative for Influenza A(H1N1)v infection, and 8 samples from Quality Control for Molecular Diagnostics (QCMD) with known viral load of types and subtypes of 17 respiratory viruses, by the Clinical Array Technology (CLARTR) PneumoVir kit® (Geomica), a 120 spots array that make possible the specific identification of: INFV A, B and C; IPV 1, 2, 3 and 4 (subtypes A and B); RSV type A (RSV-A) and type B (RSV-B); RV; MPV (subtypes A and B); ENT (Echovirus); ADV; CoV and BoV. An internal control is included, to assure that the amplification step is performed successfully and to avoid false negative results. We used a proprietary image processing software, installed in a reader (SAICLART®), that is able to detect and resolve the genotypes automatically, avoiding the subjectivity that may introduce the user interaction, and provide fast, accurate and reproducible results. All samples have been analyzed three times and all results have been confirmed by single virus RT PCR (Roche) and Light Cycler, Roche, detector. All viruses detected by PneumoVir kit in the analyzed samples have been confirmed by RT PCR (Roche), in some cases at a sensitivity level higher than what was declared from the manufacturer. We detected single infection with: RSVA, MPV A, Coronavirus 229. One coinfection of MPV A, RSV A, RSV B, one coinfection of Corona 229, IPV3, RSV A, and one coinfection of BoV, MPV A. Only in the sample with the coinfection Corona 229, IPV3, RSV A, the Real Time PCR did not confirm the presence of IPV3 genome. Therefore we believe that the CLART can readily detect respiratory viruses in various clinical respiratory samples (pharyngeal and nasal swabs, nasopharyngeal lavage, pharyngeal exudates). The signal intensity increased according to the viral titer. These data directly correlate sample viral titers with the successful detection by the CLART and highlight the importance of sensitivity when utilizing the Virochip platform in a clinical settings. The CLART positively identified respiratory viruses in the all QCMD samples randomly mixed demonstrating high specificity inside the range of the sensitivity of the method. Together, the data in this report demonstrate that the CLART can successfully detect respiratory viruses frequently found in human respiratory swabs. The ability of the CLART to positively detect viruses with a high degree of genetic variance, as is found in the specimens tested here, is a benefit that may outweigh concerns regarding costs and turnaround time. Furthermore, the advantages in the technical effort, cost, and turnaround time involved in using the CLART as a viral discovery platform far exceed those of next-generation sequencing platforms. In conclusion CLART provides, at very competitive cost, a system capable of detecting and identifying simultaneously several respiratory viruses, in clinical specimens with high sensitivity and specificity.

Author details

B. Matteoli and L. Ceccherini-Nelli

Virology Unit, Pisa University Hospital (Azienda Ospedaliero-Universitaria Pisana)
and Retrovirus Centre of the Virology Section, Department of Experimental Pathology BMIE,
University of Pisa, Pisa, Italy

4. References

[1] Nair H, Brooks WA, Katz M, Roca A, Berkley JA, Madhi SA, Simmerman JM, Gordon A, Sato M, Howie S, Krishnan A, Ope M, Lindblade KA, Carosone-Link P, Lucero M, Ochieng W, Kamimoto L, Dueger E, Bhat N, Vong S, Theodoratou E, Chittaganpitch M, Chimah O, Balmaseda A, Buchy P, Harris E, Evans V, Katayose M, Gaur B, O'Callaghan-Gordo C, Goswami D, Arvelo W, Venter M, Briese T, Tokarz R, Widdowson MA, Mounts AW, Breiman RF, Feikin DR, Klugman KP, Olsen SJ, Gessner BD, Wright PF, Rudan I, Broor S, Simões EA, Campbell H. Global burden of respiratory infections due to seasonal influenza in young children: a systematic review and meta-analysis. Lancet. 2011 Dec 3;378(9807):1917-30. Epub 2011 Nov 10. Review.

[2] Heyman PVV, Carper HT, Murphy DD, Platss-Mills TA, Patrie J, McLaughlin AP,et al. Viral infections in relation to age, atopy, and season of admission among children hospitalized for wheezing. J Allergy Clin Immunol. 2004;114: 239-47.

[3] World Health Organization. Cumulative Number of Confirmed Human Cases of Avian Influenza A/(H5N1) Reported to WHO. Available at http://www.who.int/csr/disease/avian_influenza/country/cases_table_2010_08_12/en/in dex.html. Accessed August 24, 2010.

[4] Bedford , Elliman D. Prevention, diagnosis and management of pertussis. Nursing Times. 2006. 102(46): 42–4.

[5] Sjaak de Wit JJ, Cook JK, van der Heijden HM Infectious bronchitis virus variants: a review of the history, current situation and control measures.Avian Pathol. 2011 Jun;40(3):223-35. Review.

[6] Watanabe A, Goto H, Kohno S, Matsushima T, Abe S, Aoki N, Shimokata K, Mikasa K, Niki Y.Nationwide survey on the 2005 Guidelines for the Management of Community-Acquired Adult Pneumonia: Validation of severity assessment. Respir Investig. 2012 Mar;50(1):14-22.

[7] Liaw FY, Wang CC, Chang YW, Chen SJ. Community-acquired Streptococcus Viridans Pneumonia in a Healthy Child. Indian Pediatr. 2012 Apr 8;49(4):324-6.

[8]   Okada T, Morozumi M, Sakata H, Takayanagi R, Ishiwada N, Sato Y, Oishi T, Tajima T,
      Haruta T, Kawamura N, Ouchi K, Matsubara K, Chiba N, Takahashi T, Iwata S,
      Ubukata K. A practical approach estimating etiologic agents using real-time PCR in
      pediatric inpatients with community-acquired pneumonia. J Infect Chemother. 2012
      May 9. [Epub ahead of print].

[9]   Kim L. C, Lin G, Pankuch GA, Bajaksouzian S, Jacobs MR, Appelbaum PC.
      Susceptibilities of *Haemophilus influenzae* and *Moraxella catarrhalis* to ABT-773 Compared
      to Their Susceptibilities to 11 Other Agents. Antimicrob Agents Chemother. 2001
      January; 45(1): 67–72.

[10] Kunst H, Mack D, Kon OM, Banerjee AK et al. Parasitic infections of the lung: a guide
      for the respiratory physician Thorax 2011;66:528-536. Review

[11] Saif LJ. Animal coronaviruses: what can they teach us about the severe acute respiratory
      syndrome? Rev Sci Tech. 2004 Aug;23(2):643-60.

[12] Braciale TJ, SunJ, Taeg S. Regulating the adaptive immune response to respiratory virus
      infection. Nature Reviews Immunology 2012. Apr; 12: 295-305.
      http://www.cdc.gov/flu/avian/gen-info/flu-viruses.htm.

[13] Piralla A, Pariani E, Rovida F, Campanini G, Muzzi A, Emmi V, Iotti GA, Pesenti A,
      Conaldi PG, Zanetti A, Baldanti F; and the Severe Influenza A Task Force. Segregation
      of Virulent Influenza A(H1N1) Variants in the Lower Respiratory Tract of Critically Ill
      Patients during the 2010-2011 Seasonal Epidemic. PLoS One. 2011;6(12):e28332. Epub
      2011 Dec 14.

[14] Hammond S, Chenever E, Durbin JE. Respiratory virus infection in infants and
      children. Pediatr Dev Pathol. 2007 May-Jun; 10 (3): 172-80.

[15] Renaud C, Campbell AP. Changing epidemiology of respiratory viral infections in
      hematopoietic cell transplant recipients and solid organ transplant recipients. Curr
      Opin Infect Dis. 2011 Aug;24(4):333-43. Review.

[16] Cooper RJ, Hoffman JR, Bartlett JG, et al: Principles of appropriate antibiotic use for
      acute pharyngitis in adults: Background. Ann Intern Med. 2001, 134: 509-517.

[17] http://www.cdc.gov/adenovirus/hcp/clinical-overview.html

[18] Zhenqiang Bi, Pierre BH, Formenty and Cathy E. Roth. Human bocavirus, a real
      respiratory tract pathogen. African Journal of Microbiology Research Oct, 2007; pp.051-056.

[19] 14   Paul AR, Steven OM, Monroe SS, et al., Characterization of a Novel Coronavirus
      Associated with Severe Acute Respiratory Syndrome www.sciencexpress.org / 1 May
      2003 / Page 1/ 10.1126/science.1085952.

[20] Jacques J, Moret h, MinetteD et al., Epidemiological, Molecular, and Clinical Features of
      Enterovirus Respiratory Infections in French Children between 1999 and 2005. J Clin
      Microbiol. 2008 January; 46(1): 206–213.

[21] Medina R, García-Sastrea. Influenza A viruses: new research developments. Nature
      Reviews Microbiology. Aug, 2009, 590-603.

[22] Regev L, Meningher T, Hindiyeh M, Mendelson E, Mandelboim M. Increase Human
      Metapneumovirus Mediated Morbidity following Pandemic Influenza Infection. PLoS
      One. 2012;7(4):e34750.

[23] Karron RA, Collins PL. Parainfluenza viruses. In: Fields BN, Knipe DM, Howley PM,
      eds. Fields Virology. 5th ed. Philadelphia: Lippincott-Williams & Wilkins Publishers;
      2007: 1497-1526.

[24] Papadopoulos NG, Bates PJ, Bardin PG, Papi A, Leir SH, Fraenkel DJ, Meyer J, Lackie PM, Sanderson G, Holgate ST, Johnston SL. Rhinoviruses infect the lower airways. J Infect Dis. 2000 Jun;181(6):1875-84.

[25] Respiratory Syncytial Virus". CDC, Respiratory and Enteric Viruses Branch. Reviewed on October 17, 2008. Retrieved 2009-02-10.

[26] Mahony JB. Detection of Respiratory Viruses by Molecular Methods. Clin. Microbiol. Rev. October 2008 vol. 21 no. 4 716-747.

[27] Chiu CY, Alizadeh A, Rouskin S et al. Diagnosis of a critical respiratory illness caused by human Metapneumovirus by use of a pan-virus microarray. J Clin Microbiol 2007, 45, 2340-2343.

[28] Spicuzza L, Spicuzza A, La Rosa M, Polosa R, Di Maria G. New and emerging infectious diseases. Allergy Asthma Proc. 2007; 28 (1): 28-34.28.

[29] Henrickson KJ. Cost-effective use of rapid diagnostic techniques in the treatment and prevention of viral respiratory infections. Pediatr Ann. 2005 Jan;34(1):24-31.

[30] Mahony JB. Nucleic acid amplification-based diagnosis of respiratory virus infections. Expert Rev Anti Infect Ther. 2010 Nov;8(11):1273-92.

[31] Erdman DD, Weinberg GA, Edwards KM, Walker FJ, Anderson BC, Winter J, GeneScan reverse transcription-PCR assay for detection of six common respiratory viruses in young children hospitalized with acute respiratory illness. J Clin Microbiol. 2003 Sep;41(9):4298-303.

[32] Templeton KE, Scheltinga SA, van den Eeden WC, Graffelman AW, van den Broek PJ, Claas EC. Improved diagnosis of the etiology of community-acquired pneumonia with real-time polymerase chain reaction. Clin Infect Dis. Aug 2005;41(3):345-51.

[33] Marcos MA, Camps M, Pumarola T, et al. The role of viruses in the aetiology of community-acquired pneumonia in adults. Antivir Ther. 2006;Vol. 11:351-359.

[34] Korppi M, Don M, Valent F, Canciani M. The value of clinical features in differentiating between viral, pneumococcal and atypical bacterial pneumonia in children. Acta Paediatr. Jul 2008;97(7):943-7.

[35] Jennings LC, Anderson TP, Beynon KA, Chua A, Laing RT, Werno AM, et al. Incidence and characteristics of viral community-acquired pneumonia in adults. Thorax. Jan 2008;63(1):42-8.

[36] Fox JD. Nucleic acid amplification tests for the detection and analysis of respiratory viruses: the future for diagnostics? Future Microbiol. 2007 Apr;2(2):199-211.

[37] Jartti T, Jartti L, Ruuskanen O, Söderlund-Venermo M . New respiratory viral infections. Curr Opin Pulm Med. 2012 May;18(3):271-8.

[38] Freymuth F, Vabret A, Cuvillon-Nimal D, Simon S, Dina J, Legrand L, Gouarin S, Petitjean J, Eckart P, Brouard J. Comparison of multiplex PCR assays and conventional techniques for the diagnostic of respiratory virus infections in children admitted to hospital with an acute respiratory illness. J Med Virol. 2006 Nov;78(11):1498-504.

[39] LaSala PR, Bufton KK, Ismail N, Smith MB. Prospective comparison of R-mix shell vial system with direct antigen tests and conventional cell culture for respiratory virus detection. J Clin Virol. 2007 Mar;38(3):210-6.

[40] Rangaiah SJ, Raghava VP, Srinivasan s, BadrinathS. Research Shell Vial culture Assay for the rapid diagnosis of Japanese encephalitis, West Nile and Dengue-2 viral encephalitis Virology Journal 2006, 3:2.

[41] Casiano-Colon AE, Hulbert BB, Mayer TK, Walsh EE, Falsey AR. Lack of sensitivity of rapid antigen tests for the diagnosis of respiratory syncytial virus infection in adults. J Clin Virol 2003;28:169-74.

[42] Stockton J, Stephenson I, Fleming D, Zambonm. Human Metapneumovirus as a Cause of Community-Acquired Respiratory * Illness EMERING INFECTIOUS DISEASES. Sep. 2002, 2,9.

[43] Gruteke P, Glas AS, Dierdorp M, Vreede WB, Pilon JW, Bruisten SM. Practical implementation of a multiplex PCR for acute respiratory tract infections in children. J Clin Microbiol 2004;42:5596-603.

[44] Syrmis MW, Whiley DM, Thomas M, Mackay IM, Williamson J, Siebert DJ, et al. A sensitive, specific, and cost-effective multiplex reverse-transcriptase–PCR assay for the detection of seven common respiratory viruses in respiratory samples. J Mol Diagn 2004;6:125-31.

[45] Bellau-Pujol S, Vabret A, Legrand L, Dina J, Gouarin S, Petitjean-Lecherbonnier J, Pozzetto B, Ginevra C, Freymuth F. Development of three multiplex RT-PCR assays for the detection of 12 respiratory RNA viruses. J Virol Methods 2005;126:53-63.

[46] Coiras MT, Aguilar JC, Garcia ML, Casas I, Perez-Brena P. Simultaneous detection of fourteen respiratory viruses in clinical specimens by two multiplex reversetranscription nested-PCR assays. J Med Virol 2004;72:484-95.

[47] Freymuth F, Vabret A, Cuvillon-Nimal D, Simon S, Dina J, Legrand L, et al. Comparison of multiplex PCR assays and conventional techniques for the diagnostic of respiratory virus infections in children admitted to hospital with an acute respiratory illness. J Med Virol 2006;78:1498-504.

[48] Kehl SC, Henrickson KJ, Hua W, Fan J. Evaluation of the Hexaplex assay for detection of respiratory viruses in children. J Clin Microbiol 2001;39:1696-701.

[49] Chiu CY, Rouskin S, Koshy A, Urisman A, Fischer K, Yagi S, Schnurr D, Eckburg PB, Tompkins LS, Blackburn BG, Merker JD, Patterson BK, Ganem D, DeRisi JL. Microarray detection of human parainfluenzavirus 4 infection associated with respiratory failure in an immunocompetent adult. Clin Infect Dis 2006. 43, e71-76.

[50] Lee WM, Grindle K, Pappas T, Marshall DJ, Moser MJ, Beaty EL, et al. High-throughput, sensitive, and accurate multiplex PCR-microsphere flow cytometry system for large-scale comprehensive detection of respiratory viruses. J Clin Microbiol 2007;45:2626-34.

[51] Kistler, A.,et al. Pan-viral screening of respiratory tract infections in adults with and without asthma reveals unexpected human Coronavirus and human rhinovirus diversity. J Infect Dis 2007. 196, 817-825.

[52] Wang, D.,et al. Microarray-based detection and genotyping of viral pathogens. Proc Natl Acad Sci U S A 2002. 99, 15687-15692.

[53] Brockmeier LS, Miller LC, Faaberg KS et al. Samples of Swine Respiratory Viruses in Clinical Utility of a Panviral Microarray for Detection J. Clin. Microbiol. 2011, 49(4):1542.

[54] Wang D, Urisman A, Liu YT, Springer M, Ksiazek TG, Erdman DD, et al. Viral discovery and sequence recovery using DNA microarrays. PLoS Biol 2003;1:E2.

[55] Mahony JB, Petrich A, Smieja M. Molecular diagnosis of respiratory virus infections. Crit Rev Clin Lab Sci. 2011 Sep;48(5-6):217-49.

Permissions

The contributors of this book come from diverse backgrounds, making this book a truly international effort. This book will bring forth new frontiers with its revolutionizing research information and detailed analysis of the nascent developments around the world.

We would like to thank Prof. Luca Ceccherini-Nelli, MD and Dott.ssa Barbara Matteoli, for lending their expertise to make the book truly unique. They have played a crucial role in the development of this book. Without their invaluable contribution this book wouldn't have been possible. They have made vital efforts to compile up to date information on the varied aspects of this subject to make this book a valuable addition to the collection of many professionals and students.

This book was conceptualized with the vision of imparting up-to-date information and advanced data in this field. To ensure the same, a matchless editorial board was set up. Every individual on the board went through rigorous rounds of assessment to prove their worth. After which they invested a large part of their time researching and compiling the most relevant data for our readers. Conferences and sessions were held from time to time between the editorial board and the contributing authors to present the data in the most comprehensible form. The editorial team has worked tirelessly to provide valuable and valid information to help people across the globe.

Every chapter published in this book has been scrutinized by our experts. Their significance has been extensively debated. The topics covered herein carry significant findings which will fuel the growth of the discipline. They may even be implemented as practical applications or may be referred to as a beginning point for another development. Chapters in this book were first published by InTech; hereby published with permission under the Creative Commons Attribution License or equivalent.

The editorial board has been involved in producing this book since its inception. They have spent rigorous hours researching and exploring the diverse topics which have resulted in the successful publishing of this book. They have passed on their knowledge of decades through this book. To expedite this challenging task, the publisher supported the team at every step. A small team of assistant editors was also appointed to further simplify the editing procedure and attain best results for the readers.

Our editorial team has been hand-picked from every corner of the world. Their multi-ethnicity adds dynamic inputs to the discussions which result in innovative outcomes. These outcomes are then further discussed with the researchers and contributors who give their valuable feedback and opinion regarding the same. The feedback is then collaborated with the researches and they are edited in a comprehensive manner to aid the understanding of the subject.

Apart from the editorial board, the designing team has also invested a significant amount of their time in understanding the subject and creating the most relevant covers. They scrutinized every image to scout for the most suitable representation of the subject and create an appropriate cover for the book.

The publishing team has been involved in this book since its early stages. They were actively engaged in every process, be it collecting the data, connecting with the contributors or procuring relevant information. The team has been an ardent support to the editorial, designing and production team. Their endless efforts to recruit the best for this project, has resulted in the accomplishment of this book. They are a veteran in the field of academics and their pool of knowledge is as vast as their experience in printing. Their expertise and guidance has proved useful at every step. Their uncompromising quality standards have made this book an exceptional effort. Their encouragement from time to time has been an inspiration for everyone.

The publisher and the editorial board hope that this book will prove to be a valuable piece of knowledge for researchers, students, practitioners and scholars across the globe.

List of Contributors

Zhanqiu Yang and Hai-Rong Xiong
School of Basic Medical Sciences, Wuhan University, The People's Republic of China

Shlomo Rottem and Jonathan D. Kornspan
Department of Microbiology and Molecular Genetics, IMRIC, The Hebrew University-Hadassah Medical School, Jerusalem, Israel

Nechama S. Kosower
Department of Human Molecular Genetics and Biochemistry, Sackler School of Medicine, Tel-Aviv University, Ramat-Aviv, Tel-Aviv, Israel

Satoru Kaneko and Kiyoshi Takamatsu
Reproduction Center, Gynecology, Ichikawa General Hospital, Tokyo Dental College, Sugano, Ichikawa, Chiba, Japan

Zhan-Qiu Yang and Hai-Rong Xiong
School of Basic Medical Sciences, Wuhan University, The People's Republic of China

Phuc Van Pham, Binh Thanh Vu, Nhan Lu Chinh Phan, Thuy Thanh Duong, Tue Gia Vuong and Ngoc Kim Phan
Laboratory of Stem Cell Research and Application, University of Science, Vietnam National University, Ho Chi Minh City, Vietnam

Giang Do Thuy Nguyen, Thiep Van Tran, Dung Xuan Pham and Minh Hoang Le
Ho Chi Minh City Oncology Hospital, Ho Chi Minh City, Vietnam

O.S. Sotnikov
Pavlov Institute of Physiology of the Russian Academy of Sciences, St. Petersburg, Russia

Seiji Omata
Department of Biomedical Engineering, Graduate School of Biomedical Engineering, Tohoku University, Japan

Yoshinori Sawae
Department of Mechanical Engineering, Faculty of Engineering, Kyushu University, Japan

Teruo Murakami
Advanced Biomaterials Division, Research Center for Advanced Biomechanics, Kyushu
University, Japan

João Bosco Barreto Filho
Federal University of Lavras; Veterinary Medicine Department; Lavras, MG, Brazil

Maira Souza Oliveira
Federal University of Minas Gerais; Veterinary Clinical and Surgery Department; Belo
Horizonte, MG, Brazil

John A. Lednicky
Environmental and Global Health, University of Florida, Gainesville, Florida, USA

Diane E. Wyatt
KC Bio, LLC, E. Santa Fe, Olathe, Kansas, USA

Aurora Longa Briceño
Faculty of Pharmacy, Microbiology and Parasitology Deparment, Laboratory of
Gastrointestinal and Urinary Syndromes "Lcda Luisa Vizcaya", Universidad de Los
Andes, Mérida, Venezuela

**Zulma Peña Contreras, Delsy Dávila Vera, Rosa Mendoza Briceño and Ernesto Palacios
Prü**
Center for Electron Microscopy "Dr. Ernesto Palacios Prü", Universidad de Los Andes,
Mérida, Venezuela

Paulina Godzik
National Institute of Public Health – National Institute of Hygiene, Department of
Virology, Warsaw, Poland

B. Matteoli and L. Ceccherini-Nelli
Virology Unit, Pisa University Hospital (Azienda Ospedaliero-Universitaria Pisana) and
Retrovirus Centre of the Virology Section, Department of Experimental Pathology BMIE,
University of Pisa, Pisa, Italy